T0265414

SPANNING TREE RESULTS FOR GRAPHS AND MULTIGRAPHS

A Matrix-Theoretic Approach

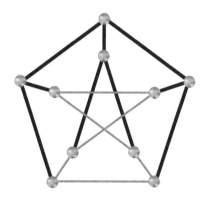

SPANNING TREE RESULTS FOR GRAPHS AND MULTIGRAPHS

A Matrix-Theoretic Approach

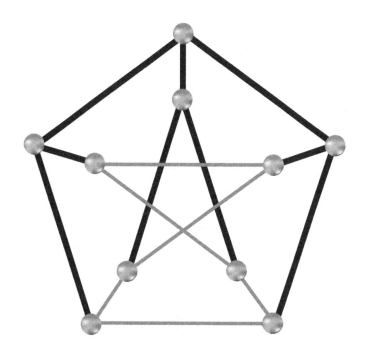

Daniel J. Gross (Seton Hall University, USA)

John T. Saccoman (Seton Hall University, USA)

Charles L. Suffel (Stevens Institute of Technology, USA)

 **World Scientific**

NEW JERSEY · LONDON · SINGAPORE · BEIJING · SHANGHAI · HONG KONG · TAIPEI · CHENNAI

Published by

World Scientific Publishing Co. Pte. Ltd.
5 Toh Tuck Link, Singapore 596224
USA office: 27 Warren Street, Suite 401-402, Hackensack, NJ 07601
UK office: 57 Shelton Street, Covent Garden, London WC2H 9HE

British Library Cataloguing-in-Publication Data
A catalogue record for this book is available from the British Library.

SPANNING TREE RESULTS FOR GRAPHS AND MULTIGRAPHS
A Matrix-Theoretic Approach

Copyright © 2015 by World Scientific Publishing Co. Pte. Ltd.

All rights reserved. This book, or parts thereof, may not be reproduced in any form or by any means, electronic or mechanical, including photocopying, recording or any information storage and retrieval system now known or to be invented, without written permission from the publisher.

For photocopying of material in this volume, please pay a copying fee through the Copyright Clearance Center, Inc., 222 Rosewood Drive, Danvers, MA 01923, USA. In this case permission to photocopy is not required from the publisher.

ISBN 978-981-4566-03-2

Printed in Singapore

This work could not have been undertaken without the influence of our departed friend and colleague, Frank T. Boesch.

To Mary, the smartest woman I know
--DJG

To John J. Saccoman, my father
--JTS

To my mother Mildred and my father Charles
--CLS

Preface

This book is concerned with the calculation of the number of spanning trees of a multigraph using algebraic and analytic techniques. We also include several results on optimizing the number of spanning trees among all multigraphs in a class, i.e., those having a specified number of nodes, n, and edges, e, denoted $\Omega(n,e)$. The problem has some practical use in network reliability theory. Some of the material in this book has appeared elsewhere in individual publications and has been collected here for the purpose of exposition. A formal probabilistic reliability model can be described as follows: the edges of a multigraph are assumed to have equal and independent probabilities of operation p, and the reliability R of a multigraph is defined to be the probability that a spanning connected subgraph operates. If ς_i denotes the number of spanning connected subgraphs having i edges, then it is easily verified that

$$R = \sum_{i=n-1}^{e} \varsigma_i p^i (1-p)^{e-i}.$$

For small values of p, the reliability polynomial is dominated by the ς_{n-1} term, and since ς_{n-1} is the number of spanning trees, graphs with a larger number of spanning trees will have greater reliability for such p. In the study of graph theory, most of the results regarding the number of spanning trees have only been proven for simple graphs, so herein, we investigate the problem for the extended class of multigraphs. It should be noted that, while extensions to multigraphs make the optimal solution readily apparent in many problems, it is not the case for the spanning tree problem.

In Chapter 0, we present some graph theory and matrix theory background material so that the reader will be familiar with the terminology used in the sequel. In Chapter 1, we introduce many algebraic results for both simple graphs and multigraphs regarding the calculation of their number of spanning trees. In Chapter 2, we present and extend a classical optimization formulation of Cheng that was useful in optimizing the number of spanning trees for certain graphs. In Chapter 3, we present a heretofore unpublished result outlined by the late Frank Boesch in the area of spanning tree enumeration of threshold graphs. In Chapter 4, we show that a complete graph minus a matching, previously shown to have the greatest number of spanning trees among all simple graphs having the same number of nodes and edges, is also optimal when the class is extended to include most multigraphs having a single multiple edge of multiplicity two. We also present an argument using degree sequences that demonstrates the optimality of this simple graph for almost all simple graphs in the class. In Chapter 5, we discuss graphs and multigraphs having all of their Laplacian eigenvalues as integers.

Contents

Chapter 0

An Introduction to Relevant
Graph Theory and Matrix Theory

This book is concerned with the calculation of the number of spanning trees of a multigraph using algebraic and analytic techniques. We also include several results on optimizing the number of spanning trees among all multigraphs in a class, i.e., those having a specified number of nodes, n, and edges, e, denoted $\Omega(n,e)$. The problem has some practical use in network reliability theory. Some of the material in this book has appeared elsewhere in individual publications and has been collected here for the purpose of exposition. In preparation, we first collect some relevant graph theoretical and matrix theoretical results.

0.1 Graph Theory

Though we assume that the reader of this work is well versed in Graph Theory, in this section we provide the primary graph theoretic definitions and operations that are used in the body of the succeeding chapters. For any other terminology and notation not provided here we refer the reader to Chartrand, Lesniak and Zhang [Chartrand, 2011].

A *multigraph* is a pair $M = (V,m)$, where m is a nonnegative integer-valued function defined on the collection of all two-element subsets of V, denoted $V_{(2)}$. In the case there are several multigraphs under consideration we use the notation $M = (V(M), m_M)$. The elements of V are called the nodes of the multigraph. We assume that V is a finite set and $n = |V|$ is the *order* of the multigraph, i.e. the order is the number of nodes. A *multiedge* is an element $\{u,v\} \in V_{(2)}$ such that $m(\{u,v\}) \neq 0$; $m(\{u,v\})$ is called the *multiplicity* of the multiedge. If for each $\{u,v\}$, $m(\{u,v\}) \in \{0,1\}$, then M is a *graph*, and $m^{-1}(\{1\}) = E$ is called the *edge set*. In this case, we use G instead of M and employ the alternate notation $G = (V, E)$.

Remark 0.1 The set of multigraphs includes the set of graphs. Thus any result that is stated for multigraphs also holds for graphs. On the other hand, results that are stated for graphs are only applicable to graphs and do not hold for multigraphs.

A multigraph has a geometric representation in which each element (node) of V is depicted by a point, and two points u and v are joined by $m(\{u,v\})$ curves. Figure 0.1 shows the geometric representation of a multigraph M and a graph G. It is traditional to refer to each point in the representation as a node and the collection of all such points as V. Also we refer to each curve in the representation as an *edge*, and the collection of all such curves as E. In the case that $m(\{u,v\}) \geq 1$ we represent any single edge between the nodes u and v by uv. The number of edges, denoted by e, is the *size* of the multigraph, i.e. $e = |E| = \sum_{\{u,v\} \in V_{(2)}} m(\{u,v\})$.

We denote the *class of all multigraphs of order n and size e* by $\Omega(n,e)$. In Figure 0.1 $M \in \Omega(5,10)$ and $G \in \Omega(5,7)$

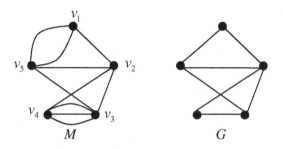

Figure 0.1: A multigraph *M* and a graph *G*.

Note that a multigraph may have several geometric representations that look different. We say that the multigraphs $M' = (V',m')$ and $M'' = (V'',m'')$ are *isomorphic* if there is a bijection $f : V' \rightarrow V''$ such that for every $\{v_1,v_2\} \in V'_{(2)}$, $m'(\{v_1,v_2\}) = m''(\{f(v_1),f(v_2)\})$. The two multigraphs depicted in Figure 0.2 are isomorphic, under the bijection $f(v_i) = u_i$.

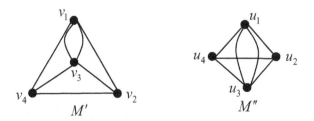

Figure 0.2: Two isomorphic multigraphs.

Edges are *incident* at its end-nodes, and two nodes which have an edge between them are called *adjacent*. Node u has *degree* deg(u) equal to the number of edges having that node as an end-node, i.e.

$$\deg(u) = \sum_{v \neq u} m(\{u,v\}) .$$

The following theorem, the First Theorem of Multigraph Theory, was proved by Euler in 1736.

Theorem 0.1 (The First Theorem of Multigraph Theory) *The sum of the degrees of all the nodes of a multigraph M is equal to twice the number of edges.*

Proof Let $x = v_i v_j$ be an edge of M, then x contributes 1 to both $\deg(v_1)$ and $\deg(v_2)$. Thus when summing the degrees of the nodes of M, each edge is counted twice. □

Another way to represent a multigraph is by a matrix. Let $M = (V, m)$ be a multigraph of order n, with node set $V = \{v_1, v_2, \ldots, v_n\}$. The *adjacency matrix* of M, denoted $\mathbf{A}(M)$ or simply $\mathbf{A}$ is the $n \times n$ matrix $[a_{ij}]$ where $a_{ij} = m(\{v_i, v_j\})$. Note if $\mathbf{A}$ is the adjacency matrix of M, then $\sum_{j=1}^{n} a_{ij} = \deg(v_i)$.

Example 0.1 The adjacency matrix of the multigraph in Figure 0.1 is

$$\mathbf{A} = \begin{bmatrix} 0 & 1 & 0 & 0 & 2 \\ 1 & 0 & 1 & 1 & 1 \\ 0 & 1 & 0 & 3 & 1 \\ 0 & 1 & 3 & 0 & 0 \\ 2 & 1 & 1 & 0 & 0 \end{bmatrix}.$$

A multigraph M is *r-regular* if each node has degree equal to r. When the specific value of r is not needed we say the multigraph is *regular*.

The multigraph $M' = (V', m')$ is called a *submultigraph* of $M = (V, m)$ if $V' \subseteq V$ and $m'(\{u, v\}) \leq m(\{u, v\})$ for all $\{u, v\} \in V'_{(2)}$. If $V' = V$, then M' is called a *spanning submultigraph*. If M is a graph then we use the terms *subgraph* and *spanning subgraph*. If $m'(\{u, v\}) = m(\{u, v\})$ for all $\{u, v\} \in V'_{(2)}$, then $M' = (V', m')$ is an *induced submultigraph* and is denoted by $\langle V' \rangle$. If $\langle V' \rangle$ contains no edges, i.e. $m(\{u, v\}) = 0$ for all $\{u, v\} \in V'_{(2)}$ then V' is called an *independent set of nodes*. If $V' = V$ and $m'(\{u, v\}) = \min\{m(\{u, v\}), 1\}$ for all $\{u, v\} \in V'_{(2)}$, then $M' = (V', m')$ is a spanning subgraph called the *underlying graph of the multigraph* (V, m). In Figure 0.1 G is the underlying graph of multigraph M.

A *path* in a multigraph M is an alternating sequence of nodes and edges $v_1, x_1, v_2, x_2, v_3, \ldots, x_{k-1}, v_k$, where $\{v_1, v_2, \ldots, v_k\}$ are distinct nodes and x_i is an edge between v_i and v_{i+1}. If the multigraph is in fact a graph then it is only necessary to list the nodes. The *length of a path* is the number of edges in the path. It is evident from the definition that a path on k nodes has length $k - 1$. A *cycle* in a multigraph M is an alternating sequence of nodes and edges $v_1,x_1,v_2,x_2,v_3,\ldots,x_{k-1},v_k,x_k,v_1$, where $\{v_0,v_1,\ldots,v_k\}$ are distinct nodes and x_i is an edge between v_{i-1} and v_i, $1\le i\le k-1$, and x_k is an edge between v_k and v_0. The *length of a cycle* is the number of edges in the cycle, so a cycle on k nodes has length k. In a multigraph it is possible to have cycles of length 2, i.e. if there are multiple edges between a pair of nodes, but in a graph the minimum cycle length is 3. A graph G is *acyclic* if it has no subgraphs which are cycles.

A multigraph is *connected* when every partition of the node set $V=V_1\cup V_2$, $V_1,V_2\ne\varnothing$, and $V_1\cap V_2=\varnothing$ has at least one multiedge with one endpoint in V_1 and the other in V_2. Alternately, a multigraph is connected if there is at least one path between every pair of nodes. A multigraph which is not connected is *disconnected*. A *component* of a multigraph M is a maximal connected submultigraph M', i.e. if M'' is a submultigraph of M that properly contains M', then M'' is disconnected. A disconnected multigraph contains at least two components, thus a multigraph is connected if and only if it has one component.

A *tree* is a connected graph with n nodes and $n-1$ edges. Considering a longest path in a tree it is easy to see that a tree has at least two nodes of degree 1, called *leaves* or *pendant* edges. A spanning subgraph that is also a tree is called a *spanning tree*. Figure 0.3 depicts a multigraph and one of its spanning trees. It is easy to see that a multigraph has spanning trees if and only if it is connected.

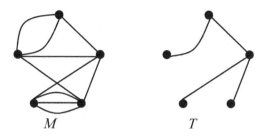

M T

Figure 0.3: A multigraph M and a spanning tree T.

One of the purposes of this book is to determine which multigraph with n nodes and e edges, i.e. in $\Omega(n,e)$, has the greatest number of spanning trees among all multigraphs in $\Omega(n,e)$. Since spanning trees exist only if the multigraph is connected, given a connected multigraph $M = (V,m)$, the following procedure will produce a spanning tree. Choose a node u at random, set $V_1 = \{u\}$, $V_2 = V - V_1$, and $E = \varnothing$. If $V_2 \neq \varnothing$ then since M is connected there is at least one edge with one endpoint in V_1 and the other in V_2. Choose one such edge, add the edge to E and remove the endpoint from set V_2 and add it to V_1. Continue this process until $V_2 = \varnothing$ and therefore $V_1 = V$. The resulting graph $T = (V_1, E)$ is a spanning tree.

The process found above can be used to find spanning trees but it is inefficient for finding all spanning trees of a multigraph; the multigraph in Figure 0.3 has 34 other spanning trees.

A *directed multigraph* is a pair $\vec{M} = (V, \vec{m})$, where $\vec{m}$ is a nonnegative integer-valued function defined on $\left(V \times V - \{(u,u) \mid u \in V\}\right)$. As for multigraphs, we refer to the elements of V as nodes. A *multi-arc is* an element $(u,v) \in (V \times V - \{(u,u) \mid u \in V\})$ such that $\vec{m}((u,v)) \neq 0$; $\vec{m}((u,v))$ is called the *multiplicity* of the multi-arc. If for each pair (u,v), $\vec{m}((u,v)) \in \{0,1\}$, then $(V,\vec{m})$ is a *directed graph* or a *digraph,* and $m^{-1}(\{1\}) = A$ is called the *arc-set.* We sometimes employ the alternate notation $\vec{D} = (V,A)$.

As in the case of multigraphs, directed multigraphs have geometric representations as well. Each element of V is depicted by a point, and two points u and v are joined by $\vec{m}((u,v))$ directed curves, directed from u to v. Figure 0.4 shows a directed multigraph and a directed graph. It is traditional to refer to each directed curve in the representation as an *arc,* and the collection of all such arcs as A. An arc from u to v is *incident from* u and *incident to* v; if such an arc exists we say u is *adjacent to* v and v is *adjacent from* u. Node u has *indegree,* indeg(u), equal to the number of arcs incident to u and *outdegree,* outdeg(u), equal to the number of arcs incident from u, i.e. $\text{indeg}(u) = \sum_{v \neq u} \vec{m}((v,u))$ and

$$\text{outdeg}(u) = \sum_{v \neq u} \vec{m}((u,v)).$$

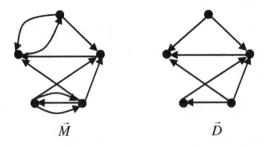

$$\vec{M} \qquad\qquad\qquad \vec{D}$$

Figure 0.4: A directed multigraph $\vec{M}$ and a directed graph $\vec{D}$.

A theorem for directed multigraphs analogous to the First Theorem of Multigraph Theory follows from the observation that each arc contributes 1 to the indegree of a node and 1 to the outdegree of another node.

Theorem 0.2 (The First Theorem of Directed Multigraph Theory)
The sum of the indegrees of all the nodes of a directed multigraph is equal to the sum of the outdegrees, and these both equal the number of arcs.

Let $\vec{M} = (V, \vec{m})$ be a directed multigraph and define $m(\{u, v\}) = \vec{m}((u, v)) + \vec{m}((v, u))$ for each two element subset $\{u, v\} \in V_{(2)}$. The multigraph $M = (V, m)$ is called the *underlying multigraph* of $\vec{M}$. Alternatively, if M is a given multigraph, then any directed multigraph having M as its underlying multigraph is referred to as an *orientation* of M.

In Figure 0.5, M is the underlying multigraph of the multi directed graph $\vec{M}$, while $\vec{M}$ can be considered an orientation of M.

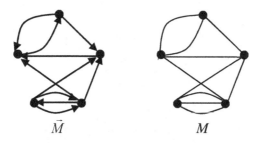

$\vec{M}$ M

Figure 0.5

We now introduce some special graphs that we will encounter in this work.

A *complete graph* is a graph of the form $(V, V_{(2)})$, i.e. there is an edge between every pair of nodes. If $|V| = n$ we denote the complete graph of order n by K_n. We depict two complete graphs in Figure 0.6.

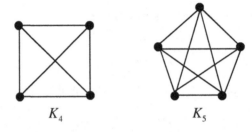

$$K_4 \qquad\qquad K_5$$

Figure 0.6: The complete graphs K_4 and K_5.

A subgraph $G' = (V', E')$ of a graph $G = (V, E)$ is a *clique* if it is a maximal complete subgraph of G, i.e. $E' = V'_{(2)}$ and for any subset of nodes V'' that properly contain V', $\langle V'' \rangle$ is not complete. In the graph depicted in Figure 0.7, $\langle \{v_1, v_2, v_3, v_4\} \rangle$ is a clique, while $\langle \{v_1, v_2, v_3\} \rangle$ is complete but not a clique.

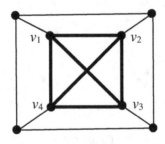

Figure 0.7: A clique inside a graph. The clique edges are thicker.

A graph $G = (V, E)$ is *bipartite* if V can be partitioned into two sets V_1 and V_2 such that any edge in E has one endpoint in V_1 and the other in V_2. A bipartite graph is a *complete bipartite graph* if for every $u \in V_1$ and every $v \in V_2$, $uv \in E$. If $|V_1| = p$ and $|V_2| = q$ we denote the complete bipartite graph by $K_{p,q}$. When $p = 1$, we refer to the graph $K_{1,q}$ as a *star*. More generally, a graph $G = (V, E)$ is *k-partite* if V can be partitioned into k sets $V_1, V_2, \ldots, V_k$, such that any edge in E has one endpoint in V_i and the other in V_j, for some $1 \le i < j \le k$. Note a 2-partite graph is bipartite. A k-partite graph is a *complete k-partite graph* if for every $u \in V_i$ and every $v \in V_j$, $uv \in E$ whenever $i \ne j$. If a complete k-partite graph is regular, then each part has the same order. If $|V_i| = n_i$ for $1 \le i \le k$, then we denote the complete k-partite graph by $K_{n_1, n_2, \ldots, n_k}$. Figure 0.8 depicts the complete bipartite graph $K_{2,3}$, the star $K_{1,4}$ and the complete 3-partite graph $K_{2,2,3}$.

$K_{2,3}$ $K_{1,4}$

Figure 0.8: *k*-partite graphs.

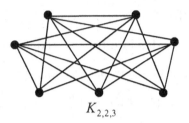

$K_{2,2,3}$

Figure 0.8: *k*-partite graphs (continued).

Suppose $n_1, n_2, \ldots, n_k$ and n are positive integers such that $n_1 < n_2 < \ldots < n_k \le \dfrac{n}{2}$. The *circulant graph* $C_n(n_1, n_2, \ldots, n_k)$ is the graph having node set $V = \{0, 1, 2, \ldots, n-1\}$ with node i adjacent to each node $(i \pm n_j) \bmod n$ for each j, $1 \le j \le k$. The $n_1, n_2, \ldots, n_k$ denote the "jump" sizes. If $n_j = j$ for all j, $1 \le j \le k$, we denote the circulant by C_n^k. Figure 0.9 shows two circulants.

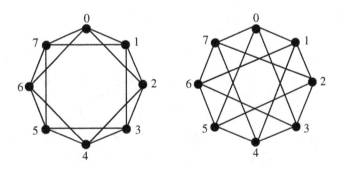

Figure 0.9: The circulants C_8^2 **and** $C_8(1,3)$.

Given a graph $G=(V,E)$, the *complement of G*, denoted $\overline{G}$, is the graph $\left(V,V_{(2)}-E\right)$, i.e. $\overline{G}$ has the same node set as G and uv is an edge of $\overline{G}$ if and only if uv is not an edge of G. A graph and its complement are shown in Figure 0.10.

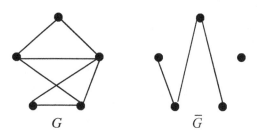

$$G \qquad \overline{G}$$

Figure 0.10: A graph and its complement.

On occasion it is necessary to delete a set of nodes or a set of edges from a multigraph. Let $M=(V,m)$ be a multigraph with edge set E. If $U \subseteq V$ is a set of nodes, then $M-U$ is the multigraph obtained by deleting the node set U and all edges incident on a node in U, i.e. $M-U=(V-U,m')$ where $m'\left(\{u,v\}\right)=m\left(\{u,v\}\right)$ for all $\{u,v\} \in (V-U)_{(2)}$. If $U=\{u\}$ we write $M-u$ instead of $M-\{u\}$. If $F \subseteq E$ is a set of edges, then, $M-F$ is the multigraph obtained by deleting the edges in F, but not the end-nodes. If x is an edge of M and $F=\{x\}$ we write $M-x$ instead of $M-\{x\}$.

Remark 0.2 There is an ambiguous case, i.e. $M-\{u,v\}$, where $\{u,v\}$ can be either the set consisting of the two nodes u and v or a multiedge incident on the two nodes u and v. If we are deleting the two node set we will write $M - u - v$ and if we are deleting an edge incident on u and v we will write $M - uv$.

Remark 0.3   One can also think of the complement of a graph $G = (V, E)$ as the graph obtained by deleting the edge set E from the complete graph on node set V.

Example 0.2   Consider the multigraph M, depicted in Figure 0.11a, with specified nodes labeled u and v and specified edges x and y. The multigraphs $M - u - v$ and $M - \{x, y\}$ are shown in Figure 0.11b and 0.11c, respectively.

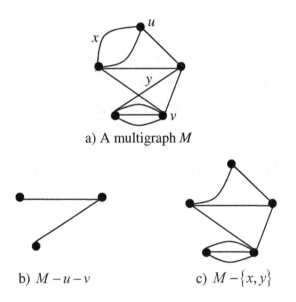

a) A multigraph M

b) $M - u - v$ c) $M - \{x, y\}$

Figure 0.11

A set of edges in a multigraph is a *matching* or an *independent set of edges* if no two edges in the set are incident at the same node. A matching consists of k edges, which we denote by kK_2, since each edge is isomorphic to K_2. We note that by the definition of a matching, $k \le \left\lfloor \dfrac{n}{2} \right\rfloor$. A special case of edge deletion that we will encounter is K_n *minus a matching*, denoted $K_n - kK_2$. In Figure 0.12 we depict $K_5 - 2K_2$. Note, the two edges deleted are arbitrary as long as they are not incident on a node, as all such graphs are isomorphic.

Figure 0.12: $K_5 - 2K_2$. **The deleted edges are indicated by the dotted edges.**

At times it may be necessary to add an edge to a multigraph. Let $M = (V, m)$ be a multigraph and $u, v \in V$, then $M^{+k \cdot uv}$ is the multigraph obtained by adding k edges between the nodes u and v, i.e.

$$M^{+k \cdot uv} = (V, m'), \text{where } m'\left(\{u', v'\}\right) = \begin{cases} m\left(\{u, v\}\right) + k & \text{if } \{u', v'\} = \{u, v\} \\ m\left(\{u', v'\}\right) & \text{otherwise} \end{cases}.$$

When $k = 1$, we write M^{+uv}. If x is an arbitrary edge, then M^{+kx} denotes the multigraph with k edges added between a pair of nodes.

Example 0.3 Consider the multigraph M, depicted in Figure 0.13a, with specified nodes labeled u, v and w. The multigraphs $M^{+2 \cdot uv}$ and

M^{+uw} are shown in Figure 0.13b and 0.13c, respectively. In Figure 0.14 we show K_5^{+x}, the placement of the edge x is arbitrary.

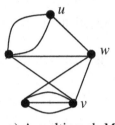

a) A multigraph M

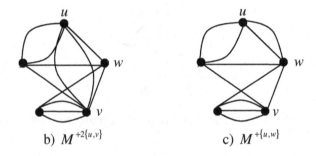

b) $M^{+2\{u,v\}}$ c) $M^{+\{u,w\}}$

Figure 0.13

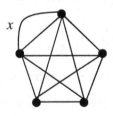

Figure 0.14: K_5^{+x}.

If $x = uv$ is an edge of a multigraph M, then the *contraction* $M \mid x$ is the multigraph obtained from coalescing u and v, the end-nodes of x, and deleting any loops which result. The contraction results in a multigraph with one fewer node and $m\big(\{u,v\}\big)$ fewer edges.

Example 0.4   Consider the multigraph M, depicted in Figure 0.15a, with specified edges labeled x and y. The multigraphs $M \mid x$ and $M \mid y$ are shown in Figure 0.15b and 0.15c, respectively.

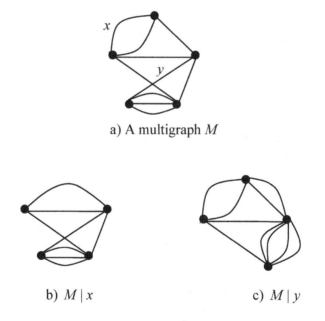

a) A multigraph M

b) $M \mid x$ c) $M \mid y$

Figure 0.15: Contractions.

We end this section by introducing some binary operations on graphs. Let $G_1 = (V_1, E_1)$ and $G_2 = (V_2, E_2)$ with $V_1 \cap V_2 = \varnothing$. The *union* of the graphs G_1 and G_2, denoted $G_1 \cup G_2$, is the graph $(V_1 \cup V_2, E_1 \cup E_2)$.

The *join* of the graphs G_1 and G_2, denoted $G_1 + G_2$, consists of $G_1 \cup G_2$ and all possible edges between the nodes in G_1 and those in G_2 (see Figure 0.16b). The *product* of the graphs G_1 and G_2, denoted $G_1 \times G_2$, is the graph $(V_1 \times V_2, E)$ where $\{(u_1, v_1), (u_2, v_2)\} \in E$ if and only if $u_1 = u_2$ and $v_1 v_2 \in E_2$ or $v_1 = v_2$ and $u_1 u_2 \in E_1$ (see Figure 0.16c).

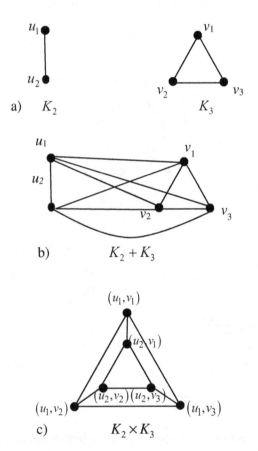

Figure 0.16: The join and the product of two graphs.

0.2   Matrix Theory

Some fundamental but perhaps less well-known results from matrix theory to be used in the sequel are included with proofs for the sake of completeness. We assume that the reader is familiar with basic matrix definitions and operations, such as size of a matrix, matrix addition, matrix multiplication, scalar multiplication transpose, and determinant. For definitions of these and any other concepts not explicitly defined herein we refer the reader to Lancaster and Tismenetsky [Lancaster, 1985].

We will first explain some matrix notation that will be used. If **P** is a matrix of size $r \times s$, then

$$\mathbf{P} = \begin{bmatrix} p_{11} & p_{12} & \cdots & p_{1s} \\ p_{21} & p_{22} & \cdots & p_{2s} \\ \vdots & \vdots & \ddots & \vdots \\ p_{r1} & p_{r2} & \cdots & p_{rs} \end{bmatrix} = \left[p_{ij} \right]_{r \times s}.$$

If the size is understood we will suppress the subscript in the latter. We refer to a matrix of size $1 \times s$ or $r \times 1$ (i.e., a row matrix or a column matrix, respectively) as a *vector*. Unless explicitly stated, all vectors are column matrices and we write

$$\mathbf{x} = \begin{bmatrix} x_1 \\ x_2 \\ \vdots \\ x_r \end{bmatrix}.$$

Finally, $\mathbf{I}_r$ denotes the $r \times r$ *identity matrix*, $\mathbf{O}_{r \times s}$ denotes the $r \times s$ matrix *zero matrix* with all entries 0, $\mathbf{0}_r$ the *zero vector* with all entries 0, and $\mathbf{1}_r$ the vector with all entries 1. If the size is understood we suppress the subscripts. We note that all matrices and vectors are **bold-faced**.

The first result we present concerns the determinant of a product of two matrices.

Theorem 0.3 – The Binèt-Cauchy Theorem [Lancaster, 1985]. *If* $\mathbf{P} = \left[p_{ij} \right]_{k \times l}$ *and* $\mathbf{Q} = \left[q_{ij} \right]_{l \times k}$ *are matrices over an arbitrary field, with* $k \leq l$, *then* $\det(\mathbf{PQ}) =$

$$\sum_{1 \leq j_1 < j_2 < \dots < j_k \leq l} \det\left(\mathbf{P} \begin{bmatrix} 1 & 2 & 3 & \dots & k \\ j_1 & j_2 & j_3 & \dots & j_k \end{bmatrix} \right) \det\left(\mathbf{Q} \begin{bmatrix} j_1 & j_2 & j_3 & \dots & j_k \\ 1 & 2 & 3 & \dots & k \end{bmatrix} \right),$$

where the minor $\mathbf{C} \begin{bmatrix} i_1 & i_2 & i_3 & \dots & i_p \\ j_1 & j_2 & j_3 & \dots & j_p \end{bmatrix}$ *denotes the submatrix of* $\mathbf{C}$ *obtained by deleting all rows except* $i_1, i_2, \dots, i_p$ *and all columns except* $j_1, j_2, \dots, j_p$. *In other words, the determinant of the product* $\mathbf{PQ}$ *is equal to the sum of the products of all possible minors of (the maximal) order k of* $\mathbf{P}$ *with corresponding minors of* $\mathbf{Q}$ *of the same order.*

Proof Observe that

$$\mathbf{PQ} = \begin{bmatrix} \sum_{\tau_1=1}^{l} p_{1\tau_1} q_{\tau_1 1} & \sum_{\tau_2=1}^{l} p_{1\tau_2} q_{\tau_2 2} & \cdots & \sum_{\tau_k=1}^{l} p_{1\tau_k} q_{\tau_k k} \\ \sum_{\tau_1=1}^{l} p_{2\tau_1} q_{\tau_1 1} & \sum_{\tau_2=1}^{l} p_{2\tau_2} q_{\tau_2 2} & \cdots & \sum_{\tau_k=1}^{l} p_{2\tau_k} q_{\tau_k k} \\ \vdots & \vdots & \ddots & \vdots \\ \sum_{\tau_1=1}^{l} p_{k\tau_1} q_{\tau_1 1} & \sum_{\tau_2=1}^{l} p_{k\tau_2} q_{\tau_2 2} & \cdots & \sum_{\tau_k=1}^{l} p_{k\tau_k} q_{\tau_k k} \end{bmatrix}.$$

Now, by successively applying the linearity of the determinant for a specific column, we obtain

$$\det(\mathbf{PQ}) = \sum_{\tau_1=1}^{l}\sum_{\tau_2=1}^{l}\cdots\sum_{\tau_k=1}^{l} \det \begin{bmatrix} p_{1\tau_1} q_{\tau_1 1} & p_{1\tau_2} q_{\tau_2 2} & \cdots & p_{1\tau_k} q_{\tau_k k} \\ p_{2\tau_1} q_{\tau_1 1} & p_{2\tau_2} q_{\tau_2 2} & \cdots & p_{2\tau_k} q_{\tau_k k} \\ \vdots & \vdots & \ddots & \vdots \\ p_{k\tau_1} q_{\tau_1 1} & p_{k\tau_2} q_{\tau_2 2} & \cdots & p_{k\tau_k} q_{\tau_k k} \end{bmatrix}$$

which is a sum of l^k determinants. Of course, if $\tau_a = \tau_b$, where $a \neq b$, then the corresponding term in the sum vanishes and the sum reduces to only those terms for which $\tau_1, \tau_2, \ldots, \tau_k$ are distinct. Now consider a specific list $1 \leq j_1 < j_2 < \ldots < j_k \leq l$, and using σ to denote permutation, it follows that $\det(\mathbf{PQ}) =$

$$\sum_{1\leq j_1 < j_2 < \ldots < j_k \leq l} \sum_{\sigma(j_1, j_2, \ldots, j_k) = (\tau_1, \tau_2, \ldots, \tau_k)} \det \begin{bmatrix} p_{1\tau_1} q_{\tau_1 1} & p_{1\tau_2} q_{\tau_2 2} & \cdots & p_{1\tau_k} q_{\tau_k k} \\ p_{2\tau_1} q_{\tau_1 1} & p_{2\tau_2} q_{\tau_2 2} & \cdots & p_{2\tau_k} q_{\tau_k k} \\ \vdots & \vdots & \ddots & \vdots \\ p_{k\tau_1} q_{\tau_1 1} & p_{k\tau_2} q_{\tau_2 2} & \cdots & p_{k\tau_k} q_{\tau_k k} \end{bmatrix}.$$

But

$$\det\begin{bmatrix} p_{1\tau_1}q_{\tau_1 1} & p_{1\tau_2}q_{\tau_2 2} & \cdots & p_{1\tau_k}q_{\tau_k k} \\ p_{2\tau_1}q_{\tau_1 1} & p_{2\tau_2}q_{\tau_2 2} & \cdots & p_{2\tau_k}q_{\tau_k k} \\ \vdots & \vdots & \ddots & \vdots \\ p_{k\tau_1}q_{\tau_1 1} & p_{k\tau_2}q_{\tau_2 2} & \cdots & p_{k\tau_k}q_{\tau_k k} \end{bmatrix} =$$

$$(-1)^{I_\sigma}\det\begin{bmatrix} p_{1j_1}q_{j_1 1} & p_{1j_2}q_{j_2 2} & \cdots & p_{1j_k}q_{j_k k} \\ p_{2j_1}q_{j_1 1} & p_{2j_2}q_{j_2 2} & \cdots & p_{2j_k}q_{j_k k} \\ \vdots & \vdots & \ddots & \vdots \\ p_{kj_1}q_{j_1 1} & p_{kj_2}q_{j_2 2} & \cdots & p_{kj_k}q_{j_k k} \end{bmatrix},$$

where I_σ denotes the number of inversions in the permutation $\sigma(j_1, j_2, \ldots, j_k) = (\tau_1, \tau_2, \ldots, \tau_k)$. Again, successively applying the linearity of determinants for a fixed column,

$$\det(\mathbf{PQ}) = \sum_{1 \le j_1 < j_2 < \ldots < j_k \le l} \quad \sum_{\sigma(j_1, j_2, \ldots, j_k)=(\tau_1, \tau_2, \ldots, \tau_k)} (-1)^{I_\sigma} q_{j_1 1}q_{j_2 2}\cdots q_{j_k k}\bullet$$

$$\det\left(\mathbf{P}\begin{bmatrix} 1 & 2 & \cdots & k \\ j_1 & j_2 & \cdots & j_k \end{bmatrix}\right).$$

Finally, the result follows from the observation that

$$\det\left(\mathbf{Q}\begin{bmatrix} 1 & 2 & \cdots & k \\ j_1 & j_2 & \cdots & j_k \end{bmatrix}\right) = \sum_{\sigma} (-1)^{I_\sigma} q_{j_1 1}q_{j_2 2}\cdots q_{j_k k}. \qquad \square$$

Example 0.5 Let $\mathbf{P} = \begin{bmatrix} 2 & -1 & 0 \\ 3 & 2 & 2 \end{bmatrix}$ and $\mathbf{Q} = \begin{bmatrix} 4 & 1 \\ 2 & -1 \\ 3 & 2 \end{bmatrix}$, then applying

Binèt-Cauchy:

$$\det(\mathbf{PQ}) = \det\begin{bmatrix} 2 & -1 \\ 3 & 2 \end{bmatrix}\det\begin{bmatrix} 4 & 1 \\ 2 & -1 \end{bmatrix} + \det\begin{bmatrix} -1 & 0 \\ 2 & 2 \end{bmatrix}\det\begin{bmatrix} 2 & -1 \\ 3 & 2 \end{bmatrix}$$

$$+ \det\begin{bmatrix} 2 & 0 \\ 3 & 2 \end{bmatrix}\det\begin{bmatrix} 4 & 1 \\ 3 & 2 \end{bmatrix} = (7)(-6)+(-2)(7)+(4)(5) = -36.$$

Multiplying the matrices, $\mathbf{PQ} = \begin{bmatrix} 6 & 3 \\ 22 & 5 \end{bmatrix}$ and $\det\begin{bmatrix} 6 & 3 \\ 22 & 5 \end{bmatrix} = -36.$

Before we proceed with the remainder of this section, we will review some pertinent material from Linear Algebra. A set of vectors $W \subseteq \mathbb{R}^n$ is a *subspace* of $\mathbb{R}^n$ if the following three conditions hold: (i) $\mathbf{0} \in W$, (ii) $\mathbf{x} + \mathbf{y} \in W$ for every $\mathbf{x}, \mathbf{y} \in W$, and (iii) $c\mathbf{x} \in W$ for every $c \in \mathbb{R}$, $\mathbf{x} \in \mathbb{R}^n$. Given a set of vectors $\{\mathbf{x}_1, \mathbf{x}_2, ..., \mathbf{x}_j\}$ the *span* of the vectors, $\mathrm{span}\{\mathbf{x}_1, \mathbf{x}_2, ..., \mathbf{x}_j\}$ is the set consisting of all linear combinations of the vectors, i.e.

$$\mathrm{span}\{\mathbf{x}_1, \mathbf{x}_2, ..., \mathbf{x}_j\} = \{c_1\mathbf{x}_1 + c_2\mathbf{x}_2 + ... + c_j\mathbf{x}_j \mid c_1, c_2, ..., c_j \in \mathbb{R}\}.$$

It is an easy exercise to show that $\mathrm{span}\{\mathbf{x}_1, \mathbf{x}_2, ..., \mathbf{x}_j\}$ is a subspace of $\mathbb{R}^n$. A set of vectors $\{\mathbf{x}_1, \mathbf{x}_2, ..., \mathbf{x}_j\}$ is linearly independent if $c_1\mathbf{x}_1 + c_2\mathbf{x}_2 + ... + c_j\mathbf{x}_j = \mathbf{0}$ only when $c_i = 0$, for all $1 \le i \le j$. If the set of vectors is not linearly independent, then it is *linearly dependent*, i.e. there exist $c_1, c_2, ..., c_j \in \mathbb{R}$ at least one of which is not 0 such that $c_1\mathbf{x}_1 + c_2\mathbf{x}_2 + ... + c_j\mathbf{x}_j = \mathbf{0}$. A *basis* of a subspace W is a linearly independent set of vectors that span W. The *dimension* of a subspace W, $\dim(W)$, is the number of vectors in any basis of W. Given two vectors

$$\mathbf{x} = \begin{bmatrix} x_1 \\ x_2 \\ \vdots \\ x_n \end{bmatrix} \text{ and } \mathbf{y} = \begin{bmatrix} y_1 \\ y_2 \\ \vdots \\ y_n \end{bmatrix}, \text{ the } dot \ product \ \mathbf{x} \cdot \mathbf{y} = \sum_{i=1}^{n} x_i y_i.$$ We will also use

the notation $\mathbf{x}^T \mathbf{y}$. Two vectors $\mathbf{x}$ and $\mathbf{y}$ are *orthogonal* if $\mathbf{x} \cdot \mathbf{y} = 0$. The *orthogonal complement* of a subspace $W \subseteq \mathbb{R}^n$, $W^\perp = \left\{ \mathbf{y} \in \mathbb{R}^n \mid \mathbf{y} \cdot \mathbf{x} = 0, \text{ for all } \mathbf{x} \in W \right\}$. The *Euclidean norm* or just *norm* of a vector, $\|\mathbf{x}\|_e$ or simply $\|\mathbf{x}\|$, is given by $\|\mathbf{x}\|_e = \sqrt{\mathbf{x} \cdot \mathbf{x}}$. A set of vectors $\left\{ \mathbf{x}_1, \mathbf{x}_2, \ldots, \mathbf{x}_j \right\}$ is *orthonormal* if the vectors each have norm 1 and are pairwise orthogonal, i.e. $\mathbf{x}_i \cdot \mathbf{x}_k = \begin{cases} 1, & i = k \\ 0, & i \neq k \end{cases}$. Any set of orthonormal vectors is necessarily linearly independent. An *orthonormal basis* is a basis whose vectors form an orthonormal set.

Let $\mathbf{A}$ be an $n \times n$ matrix. An *eigenvalue* of $\mathbf{A}$ is a scalar λ such that $\mathbf{A}\mathbf{x} = \lambda\mathbf{x}$ for some non-zero vector $\mathbf{x}$. Any non-zero vector $\mathbf{x}$ such that $\mathbf{A}\mathbf{x} = \lambda\mathbf{x}$ is an *eigenvector* associated with the eigenvalue λ. The *characteristic polynomial* of $\mathbf{A}$, $P_\mathbf{A}(\lambda) = \det(\lambda\mathbf{I} - \mathbf{A})$. Eigenvalues are the roots of the equation $P_\mathbf{A}(\lambda) = 0$.

Proposition 0.4 If $\mathbf{A}$ is an $n \times n$ real symmetric matrix then it has n real eigenvalues and an orthonormal basis of associated eigenvectors.

Proof Suppose $\mathbf{A}$ is an $n \times n$ real symmetric matrix, W is a subspace of $\mathbb{R}^n$ and W is invariant under $\mathbf{A}$ and $\mathbf{A}^T$, i.e. $\mathbf{A}(W) \subset W$ and $\mathbf{A}^T(W) \subset W$. If $\left\{ \mathbf{x}_1, \mathbf{x}_2, \ldots, \mathbf{x}_k \right\}$ is an orthonormal basis for W then there exists unique real $k \times k$ matrices $\mathbf{B}$ and $\mathbf{C}$ such that if $\mathbf{x} = \sum_{i=1}^{k} \alpha_i \mathbf{x}_i$,

$$\mathbf{Ax} = \sum_{i=1}^{k} \beta_i \mathbf{x}_i, \quad \text{and} \quad \mathbf{A}^T \mathbf{x} = \sum_{i=1}^{k} \chi_i \mathbf{x}_i, \quad \text{then} \quad \begin{bmatrix} \beta_1 \\ \vdots \\ \beta_k \end{bmatrix} = \mathbf{B} \begin{bmatrix} \alpha_1 \\ \vdots \\ \alpha_k \end{bmatrix} \quad \text{and}$$

$\begin{bmatrix} \chi_1 \\ \vdots \\ \chi_k \end{bmatrix} = \mathbf{C} \begin{bmatrix} \alpha_1 \\ \vdots \\ \alpha_k \end{bmatrix}$. It is easy to see that $\mathbf{B}^T = \mathbf{C}$ so that if $\mathbf{A}$ is symmetric

and $\mathbf{A}(W) \subset W$, then $\mathbf{A}^T(W) \subset W$ and $\mathbf{B}$ is symmetric.

Our first claim is that if $\mathbf{A}$ is symmetric then the linear operator $\mathbf{x} \to \mathbf{Ax}$ has k real eigenvalues in W (including multiplicities). Indeed, the eigenvalues of $\mathbf{B}$ are those of $\mathbf{x} \to \mathbf{Ax}$. Now $\det(\lambda \mathbf{I} - \mathbf{B})$ splits into linear factors over the complex numbers and therefore has k roots (including multiplicities) which are readily seen to be real.

Next we claim that each subspace W of dimension k which is invariant under $\mathbf{A}$ has an orthonormal basis of eigenvectors $\{\mathbf{y}_1, \mathbf{y}_2, \ldots, \mathbf{y}_k\}$ corresponding to the eigenvalues $\lambda_1, \lambda_2, \ldots, \lambda_k$. We prove this by induction on k. If $k = 1$ then W is spanned by the unit vector $\mathbf{y}_1$ and since W is invariant under $\mathbf{A}$ we have that $\mathbf{Ay}_1 = \lambda_1 \mathbf{y}_1$, i.e. λ_1 is a real eigenvalue and $\{\mathbf{y}_1\}$ is the orthonormal basis. Now suppose $2 \le k \le n$ and the claim is true for all invariant W of dimension $< k$. Let $\dim(W) = k$ and $\mathbf{A}(W) \subseteq W$. Let $\mathbf{y}_1 \in W$ be an eigenvalue of norm one associated with the real eigenvalue λ_1. Then $W_1 = \text{span}\{\mathbf{y}_1\}$ is invariant under $\mathbf{A}$. But then $W_1^{\perp} \cap W$, the collection of all vectors in W orthogonal to W_1, is also invariant under the symmetric matrix $\mathbf{A}$, i.e. if $\mathbf{x} \in W_1^{\perp}$ then

$0 = (\mathbf{y}^T \mathbf{A}) \mathbf{x} = \mathbf{y}^T (\mathbf{A}\mathbf{x})$. Since $\dim(W_1^\perp \cap W) = k-1$, there exists $k-1$ real eigenvalues and associated orthonormal eigenvectors $\mathbf{y}_2, \mathbf{y}_3, \ldots, \mathbf{y}_k$ for $\mathbf{A}$ on $W_1^\perp \cap W$, thereby establishing the claim.

Finally we may apply this claim to $k = n$ since $\mathbf{A}(\mathbb{R}^n) \subseteq \mathbb{R}^n$.   □

Theorem 0.5 (Courant-Fischer Max-Min Theorem) [Lancaster, 1985]
If the eigenvalues of the symmetric matrix $\mathbf{A}$ *are named in order of increasing value, i.e.* $\lambda_1 \leq \lambda_2 \leq \ldots \leq \lambda_n$, *then* $\lambda_j = \max_{L_j} \min_{0 \neq \mathbf{x} \in L_j} \dfrac{\mathbf{x}^T \mathbf{A}\mathbf{x}}{\mathbf{x}^T \mathbf{x}}$ *where*

L_j *varies over all* $(n-j+1)$-*dimensional subspaces of* $\mathbb{R}^n$.

To prove the Courant-Fischer Theorem, we first require a Lemma.

Lemma 0.6 *For each* $1 \leq j \leq n$ *let* L_j *denote an arbitrary* $(n-j+1)$-*dimensional subspace of* $\mathbb{R}^n$, *and consider a symmetric matrix* $\mathbf{A}$ *with eigenvalues* $\lambda_1 \leq \lambda_2 \leq \ldots \leq \lambda_n$. *Then*

$$\lambda_j \geq \min_{0 \neq \mathbf{x} \in L_j} \frac{\mathbf{x}^T \mathbf{A}\mathbf{x}}{\mathbf{x}^T \mathbf{x}} \quad \text{and} \quad \lambda_{n-j+1} \leq \max_{0 \neq \mathbf{x} \in L_j} \frac{\mathbf{x}^T \mathbf{A}\mathbf{x}}{\mathbf{x}^T \mathbf{x}}.$$

Proof (of Lemma 0.6) Let $\mathbf{x}_1, \mathbf{x}_2, \ldots, \mathbf{x}_n$ be an orthonormal system of eigenvectors for the corresponding eigenvalues $\lambda_1 \leq \lambda_2 \leq \ldots \leq \lambda_n$ of $\mathbf{A}$, i.e., $\mathbf{A}\mathbf{x}_i = \lambda_i \mathbf{x}_i$, and let $\hat{L}_j = \text{span}\{\mathbf{x}_1, \mathbf{x}_2, \ldots, \mathbf{x}_j\}$ for $j = 1, 2, \ldots, n$. Fix j, and observe that, since $\dim(L_j) + \dim(\hat{L}_j) = (n-j+1) + j > n$, there exists a nonzero vector $\mathbf{x}_0 \in \mathbb{R}^n$ belonging to both of these subspaces. In particular, since $\mathbf{x}_0 \in \hat{L}_j$, it follows that $\mathbf{x}_0 = \sum_{k=1}^{j} \alpha_k \mathbf{x}_k$ and thus,

$$\mathbf{x}_0^{T} \mathbf{A} \mathbf{x}_0 = \left(\sum_{k=1}^{j} \alpha_k \mathbf{x}_k \right)^{T} \mathbf{A} \left(\sum_{k=1}^{j} \alpha_k \mathbf{x}_k \right) = \sum_{k=1}^{j} \lambda_k |\alpha_k|^2 .$$

Hence,

$$\lambda_1 \sum_{k=1}^{j} |\alpha_k|^2 \le \mathbf{x}_0^{T} \mathbf{A} \mathbf{x}_0 \le \lambda_j \sum_{k=1}^{j} |\alpha_k|^2 , \text{ or } \lambda_1 \le \frac{\mathbf{x}_0^{T} \mathbf{A} \mathbf{x}_0}{\mathbf{x}_0^{T} \mathbf{x}_0} \le \lambda_j .$$

The first result must follow because $\mathbf{x}_0 \in L_j$, and so

$$\min_{0 \ne \mathbf{x} \in L_j} \frac{\mathbf{x}^{T} \mathbf{A} \mathbf{x}}{\mathbf{x}^{T} \mathbf{x}} \le \frac{\mathbf{x}^{T} \mathbf{A} \mathbf{x}}{\mathbf{x}_0^{T} \mathbf{x}_0} \le \lambda_j .$$

The second result is a consequence of replacing L_j by $span\{\mathbf{x}_1, \mathbf{x}_2, ..., \mathbf{x}_j\}$. □

Proof (of Courant-Fischer) Because of Lemma 0.5, it suffices to show the existence of an $(n-j+1)$-dimensional subspace such that $\min_{0 \ne \mathbf{x} \in L_j} (\mathbf{A}\mathbf{x}, \mathbf{x}) = \lambda_j$. Indeed, taking $L_j = span\{\mathbf{x}_j, \mathbf{x}_{j+1}, ..., \mathbf{x}_n\}$, we obtain such a subspace. □

A matrix $\mathbf{A}$ is *positive semi-definite* if $\mathbf{x}^{T} \mathbf{A} \mathbf{x} \ge 0$ for all vectors $\mathbf{x}$. Note: a matrix with all its principal minors non-negative is positive semi-definite. A matrix $\mathbf{A}$ is *positive definite* if $\mathbf{x}^{T} \mathbf{A} \mathbf{x} > 0$ for all vectors $\mathbf{x} \ne \mathbf{0}$.

Proposition 0.7 If $\mathbf{A}$ is a positive semi-definite real symmetric matrix, then its eigenvalues satisfy $0 \le \lambda_1 \le \lambda_2 \le \ldots \le \lambda_n$. If $\mathbf{A}$ is positive definite, then its eigenvalues satisfy $0 < \lambda_1 \le \lambda_2 \le \ldots \le \lambda_n$.

Proof Let A be a positive semi-definite real symmetric matrix. We know from Proposition 0.4 that all λ_i are real. Next consider $\mathbf{Ax} = \lambda \mathbf{x}$, where $\mathbf{x} \ne \mathbf{0}$, so that $0 \le \mathbf{x}^T \mathbf{Ax}$ and hence $\lambda \ge 0$. The result for positive definite matrices follows by replacing the inequality $0 \le \mathbf{x}^T \mathbf{Ax}$ with $0 < \mathbf{x}^T \mathbf{Ax}$. $\qquad\qquad\square$

We refer to the list of eigenvalues of a matrix as the *spectrum* of the matrix.

Corollary 0.8 *Suppose that $\mathbf{A}$ is a positive semi-definite real symmetric $n \times n$ matrix, and $\mathbf{B}$ is a real symmetric $n \times n$ matrix. Let $\alpha_1, \alpha_2, \ldots, \alpha_n$ denote the eigenvalues of $\mathbf{A} + \mathbf{B}$ and $\beta_1, \beta_2, \ldots, \beta_n$ denote the eigenvalues of $\mathbf{B}$. Then $\alpha_k \ge \beta_k$ for all $k = 1, 2, \ldots, n$.*

Proof Note that $\mathbf{x}^T (\mathbf{A} + \mathbf{B}) \mathbf{x} = \mathbf{x}^T \mathbf{Ax} + \mathbf{x}^T \mathbf{Bx} \ge \mathbf{x}^T \mathbf{Bx}$ so that

$$\alpha_n = \max_{L_k} \frac{\mathbf{x}^T (\mathbf{A} + \mathbf{B}) \mathbf{x}}{\mathbf{x}^T \mathbf{x}} \ge \max_{L_k} \frac{\mathbf{x}^T \mathbf{Bx}}{\mathbf{x}^T \mathbf{x}} = \beta_n, \text{ and}$$

$$\alpha_k = \max_{L_k} \min_{\mathbf{0} \ne \mathbf{x} \in L_k} \frac{\mathbf{x}^T (\mathbf{A} + \mathbf{B}) \mathbf{x}}{\mathbf{x}^T \mathbf{x}} \ge$$

$$\max_{L_k} \min_{\mathbf{0} \ne \mathbf{x} \in L_k} \frac{\mathbf{x}^T \mathbf{Bx}}{\mathbf{x}^T \mathbf{x}} = \min_{\substack{\|\mathbf{y}^j\|_e = 1 \\ j = k+1, \ldots n}} \max_{\substack{\|\mathbf{x}^j\|_e = 1, \mathbf{x}^T \mathbf{y}^j = 0 \\ j = k+1, \ldots n}} \mathbf{x}^T \mathbf{Bx} = \beta_k$$

for each $k = 1, 2, \ldots, n$. $\qquad\qquad\square$

The following Lemma, the contrapositive of the Levy-Desplanques-Haddamard Theorem [Lancaster, 1985], presents a useful property of singular matrices:

Theorem 0.9 (Dominant Matrix Theorem) *If* **A** *is a singular matrix, then there exists a row i such that* $\left| a_{ii} \right| \leq \sum_{j \neq i} \left| a_{ij} \right|$.

Proof If a matrix **A** is singular, then there is a non-zero vector **x** so that **Ax** = 0. Hence, $\sum_j a_{ij} x_j = 0$ for all i, or $a_{ii} x_i = -\sum_{j \neq i} a_{ij} x_j$ for all i, and $\left| a_{i,i} \right| \left| x \right|_i \leq \sum_{j \neq i} \left| a_{ij} \right| \left| x_j \right|$. Clearly, $\left| a_{i,i} \right| \left| x \right|_i \leq \left(\max_{j \neq i} \left| x_j \right| \right) \sum_{j \neq i} \left| a_{ij} \right|$ for all i. Let i now be chosen so that $\left| x_i \right|$ is the largest value of $\left| x_k \right|$ for $k = 1, 2, \ldots, n$; hence, $\max_{j \neq i} \left| x_j \right| \leq \left| x_i \right| = \max_k \left| x_k \right|$ for all k. Further, since $\mathbf{x} \neq \mathbf{0}$, $\left| x_i \right| = \max_k \left| x_k \right| > 0$. Therefore, $\left| a_{ii} \right| \leq \sum_{j \neq i} \left| a_{ij} \right|$. $\qquad\square$

Consider the matrix $\mathbf{A} = \mathbf{M} - \lambda \mathbf{I}$ which is singular for any λ, an eigenvalue of **M**. Applying Theorem 0.7 we see that there exists an i such that $\left| m_{ii} - \lambda \right| = \left| \lambda - m_{ii} \right| \leq \sum_{j \neq i} \left| m_{ij} \right|$. Now, using the second triangle inequality $\left| x \right| - \left| y \right| \leq \left| x - y \right|$, we get $\left| m_{ii} \right| - \left| \lambda \right| \leq \left| \lambda - m_{ii} \right| \leq \sum_{j \neq i} \left| m_{ij} \right|$ and $\left| \lambda \right| - \left| m_{ii} \right| \leq \left| \lambda - m_{ii} \right| \leq \sum_{j \neq i} \left| m_{ij} \right|$. Since these inequalities are true for each eigenvalue λ, and some i, it follows that

$$\min_{i}\left(\left|m_{ii}\right|-\sum_{j\neq i}\left|m_{ij}\right|\right)\leq\left|\lambda\right|_{\min}\leq\left|\lambda\right|_{\max}\leq\max_{i}\left(\left|m_{ii}\right|+\sum_{j\neq i}\left|m_{ij}\right|\right).$$

Of course, if we are dealing with positive semi-definite matrices, then $\lambda=\left|\lambda\right|$. This leads to the following corollary:

Corollary 0.10 (Gersgorin's Theorem) *Given an n-square complex matrix* $\mathbf{M}=\left[m_{ij}\right]$ *we have eigenvalue bounds*

$$\left|\lambda_{\min}\right|\geq\min_{i}\left(\left|m_{ii}-\sum_{j\neq i}\left|m_{ij}\right|\right|\right)\text{ and }\left|\lambda_{\max}\right|\geq\max_{i}\left(\left|m_{ii}-\sum_{j\neq i}\left|m_{ij}\right|\right|\right).$$

Our next result gives a relationship between eigenvalues of a square matrix to the trace and to the determinant of the matrix.

Theorem 0.11 Let $\mathbf{A}$ be an $n\times n$ matrix, with eigenvalues $\lambda_{1},\lambda_{2},\ldots,\lambda_{n}$,

then $\sum_{i=1}^{n}\lambda_{i}=\text{trace}\left(\mathbf{A}\right)$ and $\prod_{i=1}^{n}\lambda_{i}=\det\left(\mathbf{A}\right)$.

Proof Let $\mathbf{A}=\left[a_{ij}\right]$ be an $n\times n$ matrix and $P_{\mathbf{A}}\left(\lambda\right)=\det\left(\lambda\mathbf{I}-\mathbf{A}\right)$ be the characteristic polynomial of $\mathbf{A}$. Since $P_{\mathbf{A}}\left(\lambda\right)$ is a polynomial of degree n we can write it in the form in the $P_{\mathbf{A}}\left(\lambda\right)=\lambda^{n}-p_{n-1}\lambda^{n-1}+\ldots+\left(-1\right)^{n}p_{0}$. From this form we see that $P_{\mathbf{A}}\left(0\right)=\left(-1\right)^{n}p_{0}$. But plugging 0 into $P_{\mathbf{A}}\left(\lambda\right)=\det\left(\lambda\mathbf{I}-\mathbf{A}\right)$ we get $P_{\mathbf{A}}\left(0\right)=\det\left(-\mathbf{A}\right)=\left(-1\right)^{n}\det(\mathbf{A})$, thus $p_{0}=\det\left(\mathbf{A}\right)$. Now evaluating $\det\left(\lambda\mathbf{I}-\mathbf{A}\right)$ and collecting like terms we see that the coefficient of λ^{n-1}, $-p_{n-1}$, is $-\left(a_{11}+a_{22}+\ldots+a_{nn}\right)=-\text{trace}\left(\mathbf{A}\right)$. If $\lambda_{1},\lambda_{2},\ldots,\lambda_{n}$ are the eigenvalues of **A**, i.e. the roots of $P_A(\lambda) = 0$, then we can also express $P_A(\lambda)$ in the form $P_A(\lambda) = (\lambda - \lambda_1)(\lambda - \lambda_2) \cdots (\lambda - \lambda_n)$. Multiplying out we get

$$P_A(\lambda) = \lambda^n - (\lambda_1 + \lambda_2 + \ldots + \lambda_n)\lambda^{n-1} + \ldots + (-1)^n \lambda_1 \lambda_2 \cdots \lambda_n.$$

Since the coefficients of the polynomial are unique, the results follow from equating the coefficient of λ^{n-1} and the constant term in both expansions. □

The following development of some algebraic graph theoretical ideas will be useful in the sequel. If **B** and **C** are $n \times k$ and $m \times l$ matrices over an arbitrary field, then the *Kroenecker product* of **B** and **C**, **B** $\otimes$ **C**, is the $nm \times kl$ matrix with block form

$$\begin{bmatrix} b_{11}\mathbf{C} & b_{12}\mathbf{C} & \cdots & b_{1k}\mathbf{C} \\ b_{21}\mathbf{C} & b_{22}\mathbf{C} & \cdots & b_{2k}\mathbf{C} \\ \vdots & \vdots & \ddots & \vdots \\ b_{n1}\mathbf{C} & b_{n2}\mathbf{C} & \cdots & b_{nk}\mathbf{C} \end{bmatrix}.$$

The *Kroenecker sum* of an $n \times n$ matrix **B** and $m \times m$ matrix **C**, denoted **B** $\oplus$ **C**, is simply **B** $\otimes$ **I**$_m$ + **I**$_n$ $\otimes$ **C**.

Example 0.6

$$\begin{bmatrix} 2 & -1 \\ 1 & 3 \end{bmatrix} \otimes \begin{bmatrix} 1 & 0 & 3 \\ 2 & -1 & 1 \\ 2 & 3 & 2 \end{bmatrix} = \begin{bmatrix} 2\begin{bmatrix} 1 & 0 & 3 \\ 2 & -1 & 1 \\ 2 & 3 & 2 \end{bmatrix} & -1\begin{bmatrix} 1 & 0 & 3 \\ 2 & -1 & 1 \\ 2 & 3 & 2 \end{bmatrix} \\ 1\begin{bmatrix} 1 & 0 & 3 \\ 2 & -1 & 1 \\ 2 & 3 & 2 \end{bmatrix} & 3\begin{bmatrix} 1 & 0 & 3 \\ 2 & -1 & 1 \\ 2 & 3 & 2 \end{bmatrix} \end{bmatrix} =$$

$$\begin{bmatrix} 2 & 0 & 6 & -1 & 0 & -3 \\ 4 & -2 & 2 & -2 & 1 & -1 \\ 4 & 6 & 4 & -2 & -3 & -2 \\ 1 & 0 & 3 & 3 & 0 & 9 \\ 2 & -1 & 1 & 6 & -3 & 3 \\ 2 & 3 & 2 & 6 & 9 & 6 \end{bmatrix}$$

and

$$\begin{bmatrix} 2 & -1 \\ 1 & 3 \end{bmatrix} \oplus \begin{bmatrix} 1 & 0 & 3 \\ 2 & -1 & 1 \\ 2 & 3 & 2 \end{bmatrix} = \begin{bmatrix} 2 & -1 \\ 1 & 3 \end{bmatrix} \otimes \mathbf{I}_3 + \mathbf{I}_2 \otimes \begin{bmatrix} 1 & 0 & 3 \\ 2 & -1 & 1 \\ 2 & 3 & 2 \end{bmatrix} =$$

$$\begin{bmatrix} 2\mathbf{I}_3 & -1\mathbf{I}_3 \\ 1\mathbf{I}_3 & 3\mathbf{I}_3 \end{bmatrix} + \begin{bmatrix} 1\begin{bmatrix} 1 & 0 & 3 \\ 2 & -1 & 1 \\ 2 & 3 & 2 \end{bmatrix} & 0\begin{bmatrix} 1 & 0 & 3 \\ 2 & -1 & 1 \\ 2 & 3 & 2 \end{bmatrix} \\ 0\begin{bmatrix} 1 & 0 & 3 \\ 2 & -1 & 1 \\ 2 & 3 & 2 \end{bmatrix} & 1\begin{bmatrix} 1 & 0 & 3 \\ 2 & -1 & 1 \\ 2 & 3 & 2 \end{bmatrix} \end{bmatrix} =$$

$$\begin{bmatrix} 2 & 0 & 0 & -1 & 0 & 0 \\ 0 & 2 & 0 & 0 & -1 & 0 \\ 0 & 0 & 2 & 0 & 0 & -1 \\ 1 & 0 & 0 & 3 & 0 & 0 \\ 0 & 1 & 0 & 0 & 3 & 0 \\ 0 & 0 & 1 & 0 & 0 & 3 \end{bmatrix} + \begin{bmatrix} 1 & 0 & 3 & 0 & 0 & 0 \\ 2 & -1 & 1 & 0 & 0 & 0 \\ 2 & 3 & 2 & 0 & 0 & 0 \\ 0 & 0 & 0 & 1 & 0 & 3 \\ 0 & 0 & 0 & 2 & -1 & 1 \\ 0 & 0 & 0 & 2 & 3 & 2 \end{bmatrix} =$$

$$\begin{bmatrix} 3 & 0 & 3 & -1 & 0 & 0 \\ 2 & 1 & 1 & 0 & -1 & 0 \\ 2 & 3 & 4 & 0 & 0 & -1 \\ 1 & 0 & 0 & 4 & 0 & 3 \\ 0 & 1 & 0 & 2 & 2 & 1 \\ 0 & 0 & 1 & 2 & 3 & 5 \end{bmatrix}.$$

Theorem 0.12 (An Associative Law of Products) *If the products* **EF** *and* **UV** *are defined, then* $(\mathbf{EF}) \otimes (\mathbf{UV}) = (\mathbf{E} \otimes \mathbf{U})(\mathbf{F} \otimes \mathbf{V})$.

Proof Without loss of generality, suppose that **EF** is $k \times m$ and **UV** is $l \times n$. Then $\left((\mathbf{EF}) \otimes (\mathbf{UV})\right)_{ij} = (\mathbf{EF})_{\alpha\beta}(\mathbf{UV})_{pq}$ where $i = (\alpha - 1)l + p$ with $1 \le p \le l, 1 \le \alpha \le k$; and $1 \le q \le n, 1 \le \beta \le m$. On the other hand, if **E** has s columns and **U** has t columns, then

$$((\mathbf{E} \otimes \mathbf{U})(\mathbf{F} \otimes \mathbf{V})_{ij} = \sum_{z}(\mathbf{E} \otimes \mathbf{U})_{iz}(\mathbf{F} \otimes \mathbf{V})_{zj} = \sum_{z}\left(\mathbf{E}_{\alpha\gamma}\mathbf{U}_{pr}\right)\left(\mathbf{F}_{\gamma\beta}\mathbf{V}_{rq}\right),$$

where i and j are as above, and $z = (\gamma - 1)m + r$ with $1 \le \gamma \le s,\ 1 \le r \le t$.

Hence, $(\mathbf{E} \otimes \mathbf{U})(\mathbf{F} \otimes \mathbf{V})_{ij} = \sum_{\gamma=1}^{s}\left(\mathbf{E}_{\alpha\gamma}\mathbf{F}_{\gamma\beta}\right)\sum_{r=1}^{t}\left(\mathbf{U}_{pr}\mathbf{V}_{rq}\right) = (\mathbf{EF})_{\alpha\beta}(\mathbf{UV})_{pq}.\ \square$

An immediate consequence concerns the inverse of a Kroenecker product:

Corollary 0.13 (Inverse of a Kroenecker Product) *If* **P** *and* **Q** *are invertible, then so is* $\mathbf{P} \otimes \mathbf{Q}$ *and* $(\mathbf{P} \otimes \mathbf{Q})^{-1} = \mathbf{P}^{-1} \otimes \mathbf{Q}^{-1}$.

Proof   Consider $(\mathbf{P}\otimes\mathbf{Q})(\mathbf{P}^{-1}\otimes\mathbf{Q}^{-1})=\mathbf{P}\mathbf{P}^{-1}\otimes\mathbf{Q}\mathbf{Q}^{-1}=\mathbf{I}_m\otimes\mathbf{I}_n=\mathbf{I}_{mn}$,

where $\mathbf{P}$ is $m\times m$ and $\mathbf{Q}$ is $n\times n$.   □

Theorem 0.14 (Eigenvalues of a Kroenecker Polynomial)   *Let* **B** *and*
C *be square matrices, and set* $\phi(\mathbf{B};\mathbf{C})=\sum_{i,j}\alpha_{ij}\left(\mathbf{B}^i\otimes\mathbf{C}^j\right)$ *a finite sum.*

Then the eigenvalues of $\phi(\mathbf{B};\mathbf{C})$ *are given by* $\phi(\beta;\gamma)=\sum_{i,j}\alpha_{ij}\beta^i\gamma^j$,

where β *and* γ *run through the eigenvalues of* **B** *and* **C**, *respectively.*

Proof   Since the entries of **B** and **C** are complex, there exist invertible
matrices **P** and **Q** such that $\mathbf{B}_1=\mathbf{PBP}^{-1}$ and $\mathbf{C}_1=\mathbf{QCQ}^{-1}$ are upper
triangular. Of course, $\mathbf{B}_1^i=\mathbf{PB}^i\mathbf{P}^{-1}$ and $\mathbf{C}_1^j=\mathbf{QC}^j\mathbf{Q}^{-1}$ are also upper
triangular, and therefore so is each matrix $\alpha_{ij}\left(\mathbf{B}_1^i\otimes\mathbf{C}_1^j\right)$ as well as the

sum $\sum_{i,j}\alpha_{ij}\left(\mathbf{B}_1^i\otimes\mathbf{C}_1^j\right)$. Now it follows from Theorem 0.12 and Corollary

0.13 that

$$\left(\mathbf{B}_1^i\otimes\mathbf{C}_1^j\right)=\mathbf{P}^{-1}\mathbf{B}^i\mathbf{P}\otimes\mathbf{Q}^{-1}\mathbf{C}^j\mathbf{Q}=(\mathbf{P}\otimes\mathbf{Q})^{-1}\left(\mathbf{B}^i\otimes\mathbf{C}^j\right)=(\mathbf{P}\otimes\mathbf{Q})$$

and thus $\phi(\mathbf{B}_1;\mathbf{C}_1)=\sum_{i,j}\alpha_{ij}\left(\mathbf{B}_1^i\otimes\mathbf{C}_1^j\right)=(\mathbf{P}\otimes\mathbf{Q})^{-1}$ which is equal to

$$\left(\sum_{i,j}\alpha_{ij}\left(\mathbf{B}^i\otimes\mathbf{C}^j\right)\right)(\mathbf{P}\otimes\mathbf{Q})=(\mathbf{P}\otimes\mathbf{Q})^{-1}\phi(\mathbf{B};\mathbf{C})(\mathbf{P}\otimes\mathbf{Q}).$$

Therefore, $\phi(\mathbf{B}_1;\mathbf{C}_1)$ and $\phi(\mathbf{B};\mathbf{C})$ have the same eigenvalues, but $\phi(\mathbf{B}_1;\mathbf{C}_1)$ is upper triangular with diagonal entries given by $\phi(\beta;\gamma) = \sum_{i,j} \alpha_{ij} \beta^i \gamma^j$ as β and γ vary through the eigenvalues of $\mathbf{B}$ and $\mathbf{C}$, respectively. Hence, the result follows. $\qquad\square$

Corollary 0.15 *If* $\mathbf{B}$ *and* $\mathbf{C}$ *are square matrices, then the eigenvalues of* $\mathbf{B}\otimes\mathbf{C}$ *and* $\mathbf{B}\oplus\mathbf{C}$ *are given by* $\beta\gamma$ *and* $\beta + \gamma$, *respectively, as* β *and* γ *vary through the eigenvalues of* $\mathbf{B}$ *and* $\mathbf{C}$.

Proof This result follows from Theorem 0.14 after observing that $\mathbf{B}\otimes\mathbf{C}$ and $\mathbf{B}\oplus\mathbf{C}$ are obtained from $\phi_p(b;c) = bc$ and $\phi_s(b;c) = b^0 c + bc^0$, respectively, upon substitution of $\mathbf{B}$ for b and $\mathbf{C}$ for c, and the replacement of ordinary multiplication by Kroenecker product.

$\qquad\square$

Chapter 1

Calculating the Number of Spanning Trees: The Algebraic Approach

In this chapter, we explore the algebraic method of computing $t(M)$, the number of spanning trees of a multigraph M. We then apply these techniques to some special families of graphs.

A formal probabilistic reliability model, All Terminal Reliability (ATR), can be described as follows: the edges of a multigraph M are assumed to have equal and independent probabilities of operation p, and the reliability $R(M,p)$ of a multigraph is defined to be the probability that a spanning connected subgraph operates. If ς_i denotes the number of spanning connected subgraphs having i edges, then it is easily verified that

$$R(M,p) = \sum_{i=n-1}^{e} \varsigma_i p^i (1-p)^{e-i}.$$

For small values of p, the polynomial will be dominated by the ς_{n-1} term, and since ς_{n-1} is the number of spanning trees, graphs with a larger number of spanning trees will have greater reliability for such p.

In graph theory, most of the results regarding the number of spanning trees have only been proven for graphs, so in addition to an exposition of results for graphs, we will investigate the problem for the extended class of multigraphs. It should be noted that, while extensions to multigraphs make the optimal solution readily apparent in many problems, this is not the case for the spanning tree problem.

1.1   The Node-Arc Incidence Matrix

Consider a directed multigraph $\vec{M}$ with n nodes, $\{v_1, v_2, ..., v_n\}$, and e arcs, $\{\vec{x}_1, \vec{x}_2, ..., \vec{x}_e\}$. We define an $n \times e$ matrix, called the *node-arc incidence matrix* $\mathbf{S}(\vec{M}) = [s_{ij}]$, which uniquely characterizes $\vec{M}$ by

$$
s_{ij} = \begin{cases} 1 & \text{if arc } \vec{x}_j \text{ is incident to node } v_i \\ -1 & \text{if arc } \vec{x}_j \text{ is incident from node } v_i \\ 0 & \text{otherwise} \end{cases}.
$$

Example 1.1 The directed multigraph depicted in Figure 1.1 has node-arc incidence matrix

$$
S(\vec{M}) = \begin{array}{c} \\ v_1 \\ v_2 \\ v_3 \\ v_4 \end{array} \begin{array}{c} \begin{array}{ccccccc} e_1 & e_2 & e_3 & e_4 & e_5 & e_6 & e_7 \end{array} \\ \left[\begin{array}{ccccccc} -1 & 1 & 0 & 0 & 0 & 0 & -1 \\ 1 & -1 & -1 & 0 & 0 & 0 & 0 \\ 0 & 0 & 1 & -1 & 1 & 1 & 0 \\ 0 & 0 & 0 & 1 & -1 & -1 & 1 \end{array} \right] \end{array}
$$

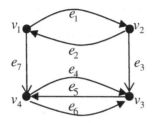

Figure 1.1: A directed multigraph.

Since each arc has exactly two end-nodes, the sum of the row vectors of $\mathbf{S}$ is $\mathbf{0}_e^T$. Clearly, any single row can be deleted from $\mathbf{S}$ without losing any information about $\vec{M}$. The matrix resulting from the deletion of the ith row is called a *reduced node-arc incidence matrix*, and is denoted $\mathbf{S}_{r(i)}\left(\vec{M}\right)$, i.e., in terms of the notation of Chapter 0, this is the submatrix

$$\mathbf{S}\begin{pmatrix} 1 & 2 & 3 & \cdots & i-1 & i+1 & \cdots & n \\ 1 & 2 & \cdots & i-1 & i & i+1 & \cdots & e \end{pmatrix}$$

or alternatively, if $\mathbf{E_i}$ denotes the matrix that results from deleting row i of the unit matrix, $\mathbf{S}_{r(i)}\left(\vec{M}\right) = \mathbf{E}_i\mathbf{S}$. Note that it is often convenient to renumber the nodes of the graph so that node n is deleted and thus node k of the renumbered nodes corresponds to row k of $\mathbf{S}_{r(n)}$.

Theorem 1.1 *The node-arc incidence matrix* $\mathbf{S}$ *is totally unimodular, i.e. every square submatrix has determinant +1, −1, or 0. Furthermore, every* $n \times n$ *submatrix has determinant* ± 1 *if and only if the* $n-1$ *selected arcs form a tree in the underlying multigraph.* [Seshu and Reed, 1961]

Proof Let $x_1, x_2, ..., x_e$ denote the edges of the underlying multigraph M, and $\mathbf{s}_1, \mathbf{s}_2, ..., \mathbf{s}_e$ denote the corresponding columns of $\mathbf{S}$. It suffices to show that $\left\{ \mathbf{s}_{j_1}, \mathbf{s}_{j_2}, ..., \mathbf{s}_{j_k} \right\}$ is linearly dependent if and only if $\left\{ x_{j_1}, x_{j_2}, ..., x_{j_k} \right\}$ (i.e., the corresponding undirected edges) contains a cycle.

(Necessity) Assume there is a directed cycle with a counterclockwise orientation. (See Figure 1.2.)

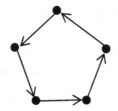

Figure 1.2: A directed cycle with counter-clockwise orientation.

Then $\sum\limits_{i=1}^{k} \alpha_i s_{j_i} = 0$ where $\alpha_i = 1$ if the jth arc is in the cycle, and $\alpha_i = 0$ otherwise. If the cycle is not counterclockwise, then we obtain a nontrivial linear combination of the columns of $\mathbf{S}$ summing to zero by multiplying those s values that correspond to arcs oriented opposite the counterclockwise direction by -1 and thereby obtaining

$$\sum_{i=1}^{k} \alpha_i s_{j_i} = 0.$$

(Sufficiency) If M is acyclic, we can add edges if necessary and direct them arbitrarily so that a tree $\vec{T}$ is formed. It suffices to show that the reduced node-arc incidence matrix for a tree, $\mathbf{S}_{r(i)}\left(\vec{T} \right)$ has determinant ± 1. The proof follows by induction on the number of nodes involved.

First assume the directed tree $\vec{T}$ has two nodes. Then $\mathbf{S}_{r(i)}\left(\vec{T}\right)$ is an 1×1 matrix with entry ± 1; hence the determinant is ± 1. Now assume that any directed tree on $n-1$ nodes has corresponding reduced node-arc incidence matrix determinant equal to ± 1. Now consider any reduced node-arc incidence matrix of a directed tree $\vec{T}$ on n nodes. As $\vec{T}$ has at least two pendant nodes, $\mathbf{S}_{r(i)}\left(\vec{T}\right)$ has at least one row having a single nonzero entry in it which, of course, corresponds to a pendant node, say v. Then expansion of $\det\left(\mathbf{S}_{r(i)}\left(\vec{T}\right)\right)$ along that row yields $\pm 1 \cdot \det\left(\mathbf{S}_{r(i)}\left(\vec{T}\right)\right)$ so, by the induction hypothesis, the result follows.   □

1.2   Laplacian Matrix

Given a multigraph M of order n, the product $\mathbf{S}\left(\vec{M}\right)\left(\mathbf{S}(\vec{M})\right)^T$ is the same for any orientation $\vec{M}$. We denote this product by $\mathbf{H}(M)$, or $\mathbf{H}$, and call it the *Laplacian* (also known as the nodal admittance matrix) of M. The terminology "nodal admittance matrix" has its origins in the theory of electrical circuit analysis.

Example 1.2 Let M be the multigraph depicted in Figure 1.3, and consider the orientation $\vec{M}$ and $S\left(\vec{M}\right)$ given in Example 1.1.

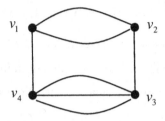

Figure 1.3

Then $\mathbf{H}(M) = \mathbf{S}(\vec{M})(\mathbf{S}(\vec{M}))^T =$

$$
\begin{bmatrix}
-1 & 1 & 0 & 0 & 0 & 0 & -1 \\
1 & -1 & -1 & 0 & 0 & 0 & 0 \\
0 & 0 & 1 & -1 & 1 & 1 & 0 \\
0 & 0 & 0 & 1 & -1 & -1 & 1
\end{bmatrix}
\begin{bmatrix}
-1 & 1 & 0 & 0 \\
1 & -1 & 0 & 0 \\
0 & -1 & 1 & 0 \\
0 & 0 & -1 & 1 \\
0 & 0 & 1 & -1 \\
0 & 0 & 1 & -1 \\
-1 & 0 & 0 & 1
\end{bmatrix} =
$$

$$
\begin{bmatrix}
3 & -2 & 0 & -1 \\
-2 & 3 & -1 & 0 \\
0 & -1 & 4 & -3 \\
-1 & 0 & -3 & 4
\end{bmatrix}.
$$

It is easily shown that $\mathbf{H}$ is the $n \times n$ matrix $\begin{bmatrix} h_{ij} \end{bmatrix}$ where

$$
h_{ij} = \begin{cases} \deg(v_i) \text{ if } i = j \\ -m(v_i, v_j) \text{ if } i \neq j \end{cases}.
$$

Note also that if $\mathbf{A}(M)$ is the adjacency matrix of M and $\mathbf{D}(M)$ is the degree matrix, i.e. a diagonal matrix whose diagonal entries are the degrees of the nodes of M, then $\mathbf{H}(M) = \mathbf{D}(M) - \mathbf{A}(M)$.

Example 1.2 continued

$$
\mathbf{H} = \begin{bmatrix} 3 & -2 & 0 & -1 \\ -2 & 3 & -1 & 0 \\ 0 & -1 & 4 & -3 \\ -1 & 0 & -3 & 4 \end{bmatrix} = \begin{bmatrix} 3 & 0 & 0 & 0 \\ 0 & 3 & 0 & 0 \\ 0 & 0 & 4 & 0 \\ 0 & 0 & 0 & 4 \end{bmatrix} - \begin{bmatrix} 0 & 2 & 0 & 1 \\ 2 & 0 & 1 & 0 \\ 0 & 1 & 0 & 3 \\ 1 & 0 & 3 & 0 \end{bmatrix} = \mathbf{D} - \mathbf{A}.
$$

The following proposition outlines some of the properties of the Laplacian matrix $\mathbf{H}(M)$ for a multigraph $M \in \Omega(n, e)$.

Proposition 1.2 *The $n \times n$ Laplacian matrix $\mathbf{H}(M)$ of the multigraph $M \in \Omega(n, e)$ has the following properties:*

(a) $\mathbf{H}(M)$ *is a real, symmetric matrix, and has real eigenvalues.*

(b) $\mathbf{H}(M)$ *is positive semi-definite, and has all nonnegative eigenvalues.*

(c) *For each i, $1 \le i \le n$, $\sum_{j=1}^{n} (\mathbf{H}(M))_{ij} = 0$.*

(d) $\mathbf{H}(M)$ *is singular, and its minimum eigenvalue is 0.*

(e) $\mathbf{H}(M)$ *is hyperdominant, i.e., $(\mathbf{H}(M))_{ii} \ge \sum_{j \ne i} |\mathbf{H}(M)|_{ij}$, for all i.*

(f) $(\mathbf{H}(M))_{ij} \le 0$, *whenever $i \ne j$.*

(g) $trace(\mathbf{H}(M)) = \sum_{i=1}^{n} (\mathbf{H}(M))_{ii} = 2e$.

Proof   (a) Follows trivially from the definitions and by Proposition 0.7.

(b) We have $\mathbf{x}^T\mathbf{H}\mathbf{x} = \mathbf{x}^T\mathbf{S}\mathbf{S}^T\mathbf{x} = \|\mathbf{x}\|_e \geq 0$. By Proposition 0.7, the eigenvalues are all nonnegative.

(c) Follows trivially from the fact $\mathbf{H} = \mathbf{D} - \mathbf{A}$.

(d) Since $\mathbf{H}\mathbf{1} = \mathbf{0}$ by (c) we have that 0 is an eigenvalue, and $\mathbf{H}$ is singular.

(e) and (f) follow trivially from the fact $\mathbf{H} = \mathbf{D} - \mathbf{A}$.

(g) Follows from the fact that $\mathbf{H} = \mathbf{D} - \mathbf{A}$ and the First Theorem of Multigraph Theory.   □

Remark 1.1   From Proposition 1.2 (a), (b) and (e) we get the eigenvalues of $\mathbf{H}(M)$ can be ordered

$$0 = \lambda_1\left(\mathbf{H}(M)\right) \leq \lambda_2\left(\mathbf{H}(M)\right) \leq \ldots \leq \lambda_n\left(\mathbf{H}(M)\right).$$

Unless otherwise stated, we assume that eigenvalues are the eigenvalues of the Laplacian $\mathbf{H}(M)$. In order to simplify notation we will use $\lambda_i(M)$ or simply λ_i when the multigraph is understood, in place of $\lambda_i\left(\mathbf{H}(M)\right)$.

We denote by $\mathbf{H}_{ij}$ the matrix resulting from deleting row i and column j from matrix $\mathbf{H}$, and the (i, j)-minor of $\mathbf{H}$ is $\det(\mathbf{H}_{ij})$. The (i, j)-cofactor is given by $C_{ij}(\mathbf{H}) = (-1)^{i+j}\det\left(\mathbf{H}_{ij}\right)$. The *adjugate matrix* of $\mathbf{H}$, adj$(\mathbf{H})$, is the transpose of the matrix formed by the cofactors of

$\mathbf{H}$, where entry (i, j) is $C_{ij}(\mathbf{H})$. Determinant computation by cofactor expansion yields $\mathbf{H}\mathrm{adj}(\mathbf{H}) = \det(\mathbf{H})\mathbf{I}$.

Next, we prove the well-known Kirchhoff's Matrix Tree Theorem [Kirchhoff, 1847]:

Theorem 1.3 *(Kirchhoff) Let $\mathbf{H}$ be the Laplacian matrix of a multigraph of order n. Then all cofactors of $\mathbf{H}$ are equal and their common value is the number of spanning trees.*

Proof First observe that Theorem 1.1 together with the Binet-Cauchy Theorem yields $t(M) = C_{ii}(\mathbf{H})$ [Brooks, 1940]. Next, observe that by Theorem 1.2, the rows of $\mathbf{S}$ sum to zero and $\mathbf{H}$ is symmetric, thus $\mathbf{H1} = \mathbf{0}$ and $\mathbf{1}^T\mathbf{H} = \mathbf{0}$, and so rank($\mathbf{H}$) is strictly less than n. We want to show that $\mathrm{adj}(\mathbf{H}) = \alpha\mathbf{J}$, where $\mathbf{J}$ is the $n \times n$ matrix of all ones and α is a scalar. If $\mathbf{H}$ is $n \times n$, with rank less than $n-1$, then $\mathrm{adj}(\mathbf{H}) = 0$, and the result is true, i.e. $\alpha = 0$. Thus, the case that remains is when the rank is $n-1$ or equivalently, the null space of $\mathbf{H}$ has dimension 1. Now $\mathbf{H}\,\mathrm{adj}(\mathbf{H}) = \mathrm{adj}(\mathbf{H})\mathbf{H} = 0\mathbf{I}$, so each column of $\mathrm{adj}(\mathbf{H})$ is an eigenvector corresponding to the eigenvalue 0 and so $\mathrm{adj}(\mathbf{H}) = \left[\alpha_1 \mathbf{1}, \alpha_2 \mathbf{1}, ..., \alpha_n \mathbf{1}\right]$ for scalars $\alpha_1, \alpha_2, ..., \alpha_n$. However, since $\mathbf{H}$ is symmetric, $\mathrm{adj}(\mathbf{H})$ is symmetric, and thus, $\left[\alpha_1 \mathbf{1}, \alpha_2 \mathbf{1}, ..., \alpha_n \mathbf{1}\right] = \begin{bmatrix} \alpha_1 \mathbf{1}^T \\ \vdots \\ \alpha_n \mathbf{1}^T \end{bmatrix}$. Hence, all α_i are equal, and $\mathrm{adj}(\mathbf{H}) = \alpha\mathbf{J}$. $\qquad\square$

Remark 1.2 A careful examination of the proof that all cofactors are equal clearly indicates that all cofactors of a matrix are equal whenever its rows and columns sum to zero, regardless of whether the matrix is symmetric. Conversely, it is easily shown from the fact $\mathbf{L}\,adj(\mathbf{L}) = adj(\mathbf{L})\,\mathbf{L} = \det(\mathbf{L})\mathbf{I}$ that every equicofactor matrix having nonzero cofactor has both row and column sum zero.

Remark 1.3 Consider $\mathbf{H} = \mathbf{SS}^{\mathbf{T}}$. Let $\mathbf{H}_{r(n)} = \mathbf{S}_{r(n)}\left(\mathbf{S}_{r(n)}\right)^{T}$ denote the *reduced Laplacian matrix*. Then Theorem 1.3 implies $\det(\mathbf{H}_{r(n)}) = t(M) = C_{ij}(\mathbf{H})$. We also note that, for a connected multigraph, $\mathbf{H}_{r(n)}$ is non-singular.

The next proposition is only applicable to graphs, as the complement of the graph G, $\overline{G}$, is not defined for multigraphs.

Proposition 1.4

(a) *If* $0 = \lambda_1(G) \le \lambda_2(G) \le \ldots \le \lambda_n(G)$ *and* $0 = \lambda_1(\overline{G}) \le \lambda_2(\overline{G}) \le \ldots \le \lambda_n(\overline{G})$ *are the eigenvalues of the Laplacian matrices* $\mathbf{H}(G)$ *and* $\mathbf{H}(\overline{G})$, *respectively, then* $\lambda_{k+2}(\overline{G}) = n - \lambda_{n-k}(G)$ *for* $k = 0, \ldots, n-2$.

(b) *The largest eigenvalue* $\lambda_n(G) = n$ *if and only if* $\overline{G}$ *is disconnected.*

(c) *[Fiedler, 1973]* $2\delta(G) - (n-2) \le \lambda_2(G) \le \kappa(G)$, *where* $\delta(G)$ *is the minimum degree of the nodes in G, and* $\kappa(G)$ *is the node connectivity of G.*

Proof (a) We begin by showing that if λ is a non-zero eigenvalue of $\mathbf{H}(G)$ with associated eigenvector $\mathbf{x}$, then $n - \lambda$ is an eigenvalue of $\mathbf{H}(\overline{G})$ is also associated with the eigenvector $\mathbf{x}$ viz. the equation

$$\mathbf{H}(\overline{G})\mathbf{x} = n\mathbf{x} - \mathbf{H}(G)\mathbf{x} = (n - \lambda)\mathbf{x}.$$

To this end, first realize that

$$\mathbf{H}(G)\mathbf{x} + \mathbf{H}(\overline{G})\mathbf{x} = \mathbf{H}(K_n)\mathbf{x} = (n\mathbf{I}_n - \mathbf{J}_n)\mathbf{x} = n\mathbf{x} - \mathbf{J}_n\mathbf{x}.$$

But $\mathbf{1}_n^T \mathbf{H}(G) = \mathbf{0}_n^T$, and $\mathbf{H}(G)\mathbf{x} = \lambda\mathbf{x}$ implies

$$(\mathbf{1}_n^T \mathbf{x}) = \mathbf{1}_n^T (\lambda\mathbf{x}) = \mathbf{1}_n^T (\mathbf{H}(G)\mathbf{x}) = \mathbf{0}_n^T \mathbf{x} = 0.$$

As $\lambda \neq 0$, it follows that $\mathbf{1}_n^T \mathbf{x} = 0$, and therefore, $\mathbf{J}_n\mathbf{x} = \mathbf{0}_n^T$. Hence, $\mathbf{H}(G)\mathbf{x} + \mathbf{H}(\overline{G})\mathbf{x} = n\mathbf{x}$, and $\mathbf{H}(\overline{G})\mathbf{x} = (n - \lambda)\mathbf{x}$ follows immediately. Now, suppose first that G is connected. Denoting the orthonormal basis associated with these eigenvalues by $\mathbf{x}_1, \mathbf{x}_2, \mathbf{x}_3, ..., \mathbf{x}_n$, it follows that $\mathbf{x}_1$ must be $\dfrac{1}{\|\mathbf{1}\|_e}$. Since $\mathbf{1}, \mathbf{x}_2, \mathbf{x}_3, ..., \mathbf{x}_n$ also correspond to the eigenvalues, $0, n - \lambda_2(G), ..., n - \lambda_n(G)$ of $\mathbf{H}(\overline{G})$, they constitute the complete set of eigenvalues of $\mathbf{H}(\overline{G})$. If, on the other hand, G is disconnected, so that $\overline{G}$ is necessarily connected, the result follows by applying the same argument to $\overline{G}$.

(b) Observe that $\overline{G}$ is disconnected if and only if $rank(\mathbf{H}(\overline{G})) \leq n - 2$. But this is the case if and only if 0 is a multiple eigenvalue of $\mathbf{H}(\overline{G})$, i.e., $0 = \lambda_2(\overline{G}) = n - \lambda_n(G)$.

(c) We defer the proof of this result until Section 1.6, where Proposition 1.19 provides a more general result.   □

Remark 1.4 Kelmans stated in an earlier publication [Kelmans, 1967] that $\lambda_2(G) \leq \delta(G)$, which is clearly subsumed by Proposition 1.4(c) since $\kappa(G) \leq \delta(G)$.

Another formula for the number of spanning trees of a multigraph involves the eigenvalues of $\mathbf{H}(M)$ [Kelmans, 1974]. Given a matrix $\mathbf{C}$ we denote by $P_{\mathbf{C}}(\lambda)$ the characteristic polynomial of $\mathbf{C}$.

Theorem 1.5: *The number of spanning trees of a multigraph is related to the eigenvalues of its Laplacian matrix as follows:*

$$t(M) = \frac{1}{n}\prod_{i=2}^{n}\lambda_i(M), \ \ where \ \ 0 = \lambda_1(M) \leq \lambda_2(M) \leq ... \leq \lambda_n(M) \ .$$

Proof Let $P_{\mathbf{H}(M)}(\lambda) = \det(\lambda\mathbf{I}_n - \mathbf{H}(M))$, the characteristic polynomial of $\mathbf{H}(M)$. The theorem is established by showing that both sides of the equation in the statement equal $\dfrac{(-1)^{n-1}P'_{\mathbf{H}(M)}(0)}{n}$.

Of course, $P_{\mathbf{H}(M)}(\lambda) = \lambda\prod_{i=2}^{n}(\lambda - \lambda_i(M))$ so that

$$P'_{\mathbf{H}(M)}(0) = \prod_{i=2}^{n}(-\lambda_i(M)) = (-1)^{n-1}\prod_{i=2}^{n}(\lambda_i(M)).$$

On the other hand, we may differentiate $\det(\lambda\mathbf{I}_n - \mathbf{H}(M))$ by adding the determinants of the matrices obtained by differentiating a single row. A typical determinant in the sum has i^{th} row consisting of a one in the ith position and other entries equal to 0; the remaining rows are those of

$\lambda \mathbf{I}_n - \mathbf{H}(M)$. Thus, when λ is set equal to 0, we obtain the determinant of $-\mathbf{H}_i(M)$, i.e., $-\mathbf{H}(M)$ with i^{th} row and column deleted. Now, an application of the Matrix Tree Theorem (Theorem 1.3) yields the conclusion that $P'_{\mathbf{H}(M)}(0) = n(-1)^{n-1} t(M)$ and the result follows.   □

Example 1.3 Consider the multigraph M given in Figure 1.4. We will compute $t(M)$ two ways, by Theorem 1.3 and by Theorem 1.5.

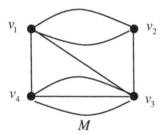

Figure 1.4

Its Laplacian is $\mathbf{H}(M) = \begin{bmatrix} 4 & -2 & -1 & -1 \\ -2 & 3 & -1 & 0 \\ -1 & -1 & 5 & -3 \\ -1 & 0 & -3 & 4 \end{bmatrix}$. Using the (1,1)-cofactor

$t(M) = (-1)^{1+1} \det \begin{bmatrix} 3 & -1 & 0 \\ -1 & 5 & -3 \\ 0 & -3 & 4 \end{bmatrix} = 29$. The eigenvalues of $\mathbf{H}(M)$ are

0, 2.676595722, 5.442973632, and 7.681330643. Thus

$$t(M) = \frac{2.676595722 \cdot 5.442973632 \cdot 7.681330643}{4} = 29.$$

We leave it to the reader to identify the 29 spanning trees.

Remark 1.5 The presence of more than one eigenvalue equal to zero implies a disconnected multigraph. In fact it is easily shown by examining the block form of the Laplacian that the number of zero eigenvalues is the same as the number of components. Thus, the formula is valid for disconnected multigraphs as well.

1.3   Special Graphs

We are now in position to find the number of spanning trees for several different families of graphs.

Proposition 1.6

(a) *(Cayley's Theorem)* [Cayley, 1889] The *number of spanning trees of K_n is n^{n-2}.*

(b) *The number of spanning trees of a graph that is regular of degree r and having adjacency matrix eigenvalues*

$$\alpha_1 \geq \alpha_2 \geq \ldots \geq \alpha_n \ \text{is} \ \frac{1}{n}\prod_{i=2}^{n}\left(r - \alpha_i\left(G\right)\right).$$

Proof (a) Since $\mathbf{H}\left(K_n\right) = n\mathbf{I}_n - \mathbf{J}_n,$ it can be shown that

$P_{\mathbf{H}(K_n)}\left(\lambda\right) = \lambda\left(\lambda - n\right)^{n-1}.$ By Theorem 1.5, $t\left(K_n\right) = \frac{1}{n}\left(n\right)^{n-1} = n^{n-2}.$

(b) As previously noted, $\mathbf{H}(G) = \mathbf{D}(G) - \mathbf{A}(G)$. Since $\mathbf{A}(G)$ is a real, symmetric matrix, then by Proposition 0.4, there exists an orthonormal basis of eigenvectors associated with the eigenvalues of $\mathbf{A}(G)$. If G is a regular graph of degree r, then it is easy to show that the eigenvectors of $\mathbf{A}(G)$ are also the eigenvectors associated with $\mathbf{H}(G)$, and the eigenvalues are related by $\lambda_i(G) = r - \alpha_i(G)$ for $i = 2, 3, ..., n$, where $r \geq \alpha_1(G) \geq ... \geq \alpha_n(G)$ are the eigenvalues of $\mathbf{A}(G)$. Application of Theorem 1.5 yields the result.   □

Our next task is to determine the number of spanning trees for the regular graph $K_n \times K_m$ using Proposition 1.6(b). To do this, we first require a result for the eigenvalues of the adjacency matrix of the product of two graphs based on the eigenvalues of the individual graphs.

Lemma 1.7   *If the nodes of G_1 and G_2 are labeled $u_1, u_2, ..., u_n$ and $v_1, v_2, ..., v_m$, respectively, and the nodes of $G_1 \times G_2$ are ordered lexicographically (i.e., label of (u_i, v_j) is smaller than that of (u_k, v_ℓ) if and only if $i < k$ or $i = k, j < \ell$,), then*

$$\mathbf{A}(G_1 \times G_2) = \mathbf{A}(G_1) \oplus \mathbf{A}(G_2).$$

Proof   Recall from Chapter 0 the Kroenecker sum:

$\mathbf{A}(G_1) \oplus \mathbf{A}(G_2) = \mathbf{I}_n \otimes \mathbf{A}(G_2) + \mathbf{A}(G_1) \otimes \mathbf{I}_m$ and consider $\mathbf{A}(G_1) \oplus \mathbf{A}(G_2)$ as an $n \times n$ matrix of blocks $\mathbf{B}_{\chi\beta}, 1 \leq \chi, \beta \leq n$, i.e.

$$\mathbf{A}(G_1) \oplus \mathbf{A}(G_2) = \begin{bmatrix} B_{11} & B_{12} & \cdots & B_{1n} \\ B_{21} & B_{22} & \cdots & B_{2n} \\ \vdots & \vdots & \ddots & \vdots \\ B_{n1} & B_{n2} & \cdots & B_{nn} \end{bmatrix}.$$

Since the $\chi\chi$ block of $I_n \otimes \mathbf{A}(G_2)$ is $\mathbf{A}(G_2)$ and the $\chi\chi$ block of

$\mathbf{A}(G_1) \otimes I_m$ is $\left(\mathbf{A}(G_1)\right)_{\chi\chi} \cdot I_m = 0$, it follows that $\mathbf{B}_{\chi\chi} = \mathbf{A}(G_2)$. Thus,

the entry of $\mathbf{A}(G_1) \oplus \mathbf{A}(G_2)$ in the row corresponding to node $\left(u_\chi, v_i\right)$

and the column corresponding to $\left(u_\chi, v_j\right)$ is $\left(\mathbf{A}(G_2)\right)_{ij}$. In other words,

the entry is 1 if v_i and v_j are adjacent in G_2 and 0 otherwise. Similarly,

if $\chi \neq \beta$, it is easily seen that $\mathbf{B}_{\chi\beta} = \left(\mathbf{A}(G_1)\right)_{\chi\beta} \cdot I_m$; thus the entry of

$\mathbf{A}(G_1) \otimes \mathbf{A}(G_2)$ in the row corresponding to $\left(u_\chi, v_i\right)$ and column

corresponding to $\left(u_\beta, v_j\right)$ is

$$\begin{cases} 0 & \text{if } i \neq j \text{ or if } i = j \text{ but } \{u_\chi, u_\beta\} \notin E_1 \\ 1 & \text{if } i = j \text{ and } \{u_\chi, u_\beta\} \in E_1 \end{cases}.$$

In conclusion, $\mathbf{A}(G_1) \oplus \mathbf{A}(G_2)$ is the adjacency matrix of $G_1 \times G_2$.   □

Corollary 1.8 *The number of spanning trees of a product of complete*

graphs $K_n \times K_m$ *is* $(m+n)^{(n-1)(m-1)} m^{m-2} n^{n-2}$.

Proof By Lemma 1.7, $\mathbf{A}(K_n \times K_m) = \mathbf{A}(K_n) \oplus \mathbf{A}(K_m)$. Since the

eigenvalues of a Kroenecker sum are obtained by summing those of the summands, we require the eigenvalues of $\mathbf{A}\left(K_p\right)$. But the eigenvalues of $\mathbf{H}\left(K_p\right)$ are p with multiplicity $p-1$, and 0 with multiplicity 1, by Proposition 1.6(a), so $\mathbf{A}\left(K_p\right)$ has eigenvalues -1 with multiplicity $p-1$ and $p-1$ with multiplicity 1. It follows that the eigenvalues of $\mathbf{A}\left(K_n \times K_m\right)$ are:

$$\alpha\left(K_n \times K_m\right) = \begin{cases} -2 & \text{with multplicity } (n-1)(m-1) \\ n-2 & \text{with multiplicity } m-1 \\ m-2 & \text{with multiplicity } n-1 \\ m+n-2 & \text{with multiplicity } 1 \end{cases}.$$

Finally, application of Proposition 1.6(b) yields

$$t\left(K_n \times K_m\right)$$
$$= \frac{1}{mn}\left(m+n-2-(-2)\right)^{(n-1)(m-1)}\left(m+n-2-(n-2)\right)^{m-1}\left(m+n-2-(m-2)\right)^{n-1}$$

and so $t(K_n \times K_m) = (m+n)^{(n-1)(m-1)} m^{m-2} n^{n-2}$. $\qquad\square$

Recall the definition of a circulant graph given in Chapter 0. For example, C_8^2 (see Figure 0.9) has adjacency matrix

$$\begin{bmatrix} 0 & 1 & 1 & 0 & 0 & 0 & 1 & 1 \\ 1 & 0 & 1 & 1 & 0 & 0 & 0 & 1 \\ 1 & 1 & 0 & 1 & 1 & 0 & 0 & 0 \\ 0 & 1 & 1 & 0 & 1 & 1 & 0 & 0 \\ 0 & 0 & 1 & 1 & 0 & 1 & 1 & 0 \\ 0 & 0 & 0 & 1 & 1 & 0 & 1 & 1 \\ 1 & 0 & 0 & 0 & 1 & 1 & 0 & 1 \\ 1 & 1 & 0 & 0 & 0 & 1 & 1 & 0 \end{bmatrix}.$$

Its rows are successive shifts to the right of the row vector $\begin{bmatrix} 0 & 1 & 1 & 0 & 0 & 0 & 1 & 1 \end{bmatrix}$. This is true in general for any circulant graph, i.e. the adjacency matrix is obtained by successively shifting the first row vector to the right by one place. More formally, row k of the circulant matrix $\mathbf{A}$ generated by the vector $\mathbf{a} = \begin{bmatrix} a_1, ..., a_n \end{bmatrix}$ is $\varepsilon^{k-1}\left(\begin{bmatrix} a_{0,...,} a_{n-1} \end{bmatrix}\right)$, where ε represents the shift operator, i.e., $\varepsilon\left(\begin{bmatrix} b_1, ..., b_n \end{bmatrix}\right) = \left(\begin{bmatrix} b_n, b_1, ..., b_{n-1} \end{bmatrix}\right)$. Furthermore, by setting $\mathbf{E}_n$ equal to the $n \times n$ circulant matrix generated by $\begin{bmatrix} 0 & 1 & 0 & . & . & . & 0 \end{bmatrix}$, i.e.

$$\mathbf{E}_n = \begin{bmatrix} 0 & 1 & 0 & 0 & \cdots & 0 \\ 0 & 0 & 1 & 0 & \cdots & 0 \\ 0 & 0 & 0 & 1 & \cdots & 0 \\ \vdots & \vdots & \vdots & \vdots & \ddots & \vdots \\ 0 & 0 & 0 & 0 & \cdots & 1 \\ 1 & 0 & 0 & 0 & \cdots & 0 \end{bmatrix}$$

we obtain $\mathbf{A} = \sum_{i=1}^{n} a_i \mathbf{E}_n^i$ where $\mathbf{E}_n^i = \left(\mathbf{E}_n\right)^i$. Those nonzero circulant matrices which are the adjacency matrices of circulant graphs are distinguished by the fact that $a_0 = 0$, $a_i = 0$, 1 for $i \leq 1$ and $a_i = 1$ if and only if $a_{n-i} = 1$, i.e., $\mathbf{A}$ is symmetric. Observe that $\mathbf{A}\left(C_8^2\right) = \mathbf{E}_8 + \mathbf{E}_8^2 + \mathbf{E}_8^6 + \mathbf{E}_8^7$. To compute the eigenvalues of a circulant matrix,

$$\det\left(\lambda \mathbf{I}_n - \mathbf{E}_n\right) =$$

$$\det \begin{bmatrix} \lambda & -1 & 0 & 0 & \cdots & 0 \\ 0 & \lambda & -1 & 0 & \cdots & 0 \\ 0 & 0 & \lambda & -1 & \cdots & 0 \\ \vdots & \vdots & \vdots & \vdots & \ddots & \vdots \\ 0 & 0 & 0 & 0 & \cdots & -1 \\ -1 & 0 & 0 & 0 & \cdots & \lambda \end{bmatrix} = \lambda \cdot \lambda^{n-1} - 1(-1)^{1+n}(-1)^{n-1} = \lambda^n - 1$$

so that the eigenvalues of $\mathbf{E}_n$ are the nth roots of unity. Hence, if $n_k < \dfrac{n}{2}$, the eigenvalues of $\mathbf{A}\big(C_n(n_1, n_2, ..., n_k)\big) = \displaystyle\sum_{i=1}^{k}\big(\mathbf{E}_n^{n_i} + \mathbf{E}_n^{n-n_i}\big)$ are given by

$$\sum_{i=1}^{k}\left(\left(e^{\frac{i2\pi j}{n}}\right)^{n_i} + \left(e^{\frac{i2\pi j}{n}}\right)^{-n_i}\right) = 2\sum_{i=1}^{k}\cos\left(\frac{2\pi n_i j}{n}\right)^{n_i} \quad \text{for } j = 0,1,...,n-1.$$

On the other hand, if $n_k = \dfrac{n}{2}$, the eigenvalues of

$$\mathbf{A}\left(C_n\left(n_1, n_2, ..., n_{k-1}, \frac{n}{2}\right)\right) = \sum_{i=1}^{k}\big(\mathbf{E}_n^{n_i} + \mathbf{E}_n^{n-n_i}\big) + \mathbf{E}_n^{\frac{n}{2}}$$

are given by

$$2\sum_{i=1}^{k}\cos\left(\frac{2\pi n_i j}{n}\right) + (-1)^j, \ j = 0,1,...,n-1 \ .$$

The preceding discussion proves the following proposition:

Proposition 1.9 *The number of spanning trees of a circulant is*

$$t\big(C_n(n_1,...,n_k)\big) =$$

$$\left[\frac{2^{n-1}}{n}\prod_{j=1}^{n-1}\left[k-\sum_{i=1}^{k}\cos\left(\frac{2\pi n_i j}{n}\right)\right]\right] \qquad \text{if } n_k < \frac{n}{2}$$

$$\left[\frac{2^{n-1}}{n}\prod_{m=1}^{\frac{n}{2}-1}\left[k-\sum_{i=1}^{k}\cos\left(\frac{4\pi n_i m}{n}\right)\right]\prod_{m=1}^{\frac{n}{2}}\left[k-1-\sum_{i=1}^{k}\cos\left(\frac{2(2m-1)\pi n_i}{n}\right)\right]\right] \qquad \text{if } n_k = \frac{n}{2}$$

$\square$

The next result, which follows from Theorem 1.5, gives the number of spanning trees for G in terms of the Laplacian matrix of the complement $\overline{G}$ of G.

Corollary 1.10 *If G is a graph on n nodes with complement $\overline{G}$ and $P_{\mathbf{H}(\overline{G})}(\lambda)$ is the characteristic polynomial of $\mathbf{H}(\overline{G})$, the Laplacian matrix of $\overline{G}$, then $t(G) = \frac{1}{n^2}P_{\mathbf{H}(\overline{G})}(n)$.*

Proof Applying Theorem 1.5, $t(G) = \frac{1}{n}\prod_{i=2}^{n}\lambda_i(G)$. By Proposition 1.4,

$\lambda_i(\overline{G}) = n - \lambda_{n-i}(G)$ for $i = 1,...,n-1$. Thus, $t(G) = \frac{1}{n}\prod_{i=2}^{n}(n - \lambda_i(\overline{G}))$.

Now, $P_{\mathbf{H}(G)}(\lambda) = \lambda\prod_{i=2}^{n}(\lambda - \lambda_i(G))$ so that $\frac{1}{n}P_{\mathbf{H}(\overline{G})}(n) = \prod_{i=2}^{n}(n - \lambda_i(G))$,

and the result follows. $\square$

This theorem can be refined after observing that, whenever $\overline{G}$ can be decomposed into a node-disjoint union of subgraphs

$C_1, C_2, ..., C_k,$ then $P_{\mathbf{H}(\bar{G})}(\lambda) = \prod_{i=1}^{k} P_{\mathbf{H}(C_i)}(\lambda).$ We record this refinement in the next result:

Corollary 1.11 *If the complement $\bar{G}$ of G is the node disjoint union of subgraphs $C_1, C_2, ..., C_k,$ then $t(G)$* $= \dfrac{1}{n^2} \prod_{i=1}^{k} P_{\mathbf{H}(C_i)}(n).$ *In particular, if G has $n - \chi$ isolated nodes, and the deletion of these nodes leaves $\bar{G}_\chi,$ then $t(G) = n^{n-\chi-2} P_{\mathbf{H}(\bar{G}_\chi)}(n).$* □

The use of the graph complement makes it easier to calculate the number of spanning trees for many types of graphs. One such example is "K_n minus a matching of size k", i.e., $K_n - kK_2$. The formula for the number of spanning trees for this and some other interesting graphs are established in the following proposition:

Proposition 1.12 *The formulas for the number of spanning trees for some special types of graphs are as follows*:

   (a) K_n minus a matching (Figure 1.5): $t(K_n - kK_2) = n^{n-2}\left(1 - \dfrac{2}{n}\right)^k.$

   (b) k-partite graphs (Figure 1.6)

      (i)   complete bipartite graph: $t(K_{p,q}) = q^{p-1} p^{q-1};$

      (ii) regular complete k-partite graph:

$$t\left(\underbrace{K_{r,r,...,r}}_{k \text{ terms}}\right) = r^{rk-2} k^{k-2} (k-1)^{(r-1)k};$$

(iii) complete bipartite with equal part sizes minus a full matching: $t\left(K_{p,p} - pK_2\right) = (p-1)\,p^{p-2}\,(p-2)^{p-1}$.

(c)

(i) complete graph of order n minus a path of order p, $n > p$. (Figure 1.7):

$$t\left(K_n - P_p\right) =$$

$$\begin{cases} \dfrac{n^{n-p-2}}{2\sqrt{\left(\dfrac{n-2}{2}\right)^2 - 1}}\left[\left(\dfrac{n-2}{2} + \sqrt{\left(\dfrac{n-2}{2}\right)^2 - 1}\right)^{p+1} - \left(\dfrac{n-2}{2} - \sqrt{\left(\dfrac{n-2}{2}\right)^2 - 1}\right)^{p+1}\right] & \text{if } n > 4 \\[4ex] 4^{3-p} \cdot p & \text{if } n = 4 \end{cases}$$

(ii) complete graph of order n minus a cycle of order p (Figure 1.8):

$$t\left(K_n - C_p\right) =$$

$$\begin{cases} n^{n-p-2}\left[\left(\dfrac{n-2}{2} + \sqrt{\left(\dfrac{n-2}{2}\right)^2 - 1}\right)^{p} + \left(\dfrac{n-2}{2} - \sqrt{\left(\dfrac{n-2}{2}\right)^2 - 1}\right)^{p} + 2(-1)^{p+1}\right] & \text{if } n > 4 \\[4ex] \dfrac{2 - 2(-1)^p}{4} & \text{if } n = 4 \text{ and } p = 3,4 \end{cases}$$

Proof (a) Suppose that E is a matching in K_n consisting of k node disjoint edges and $G = K_n - E$. Then $\bar{G} = kK_2 \cup (n-2k)K_1$, a node disjoint union. Since the characteristic polynomial of $\mathbf{H}\left(K_2\right)$ is $\lambda(\lambda - 2)$, the formula follows by Corollary 1.11.

Figure 1.5: $K_6 - 2K_2$. **The dashed edges indicate the** $2K_2$.

(b) Consider the most general complete k-partite graph G with complement $\overline{G} = \bigcup_{i=1}^{k} K_{\chi_i} \cup (n - \chi) K_1$ where the union is node-disjoint. Two special cases are the complete bipartite case and the complete k-partite case where all parts have the same order. In the first case, $k = 2$ and $n = \chi$. Setting $\chi_1 = p$ and $\chi_2 = q$, we see that

$$t\left(K_{p,q}\right) = q^{p-1} p^{q-1}.$$

In the second case, $n = \chi$ and $\chi_1 = \chi_2 = \cdots = \chi_k = r$. Thus,

$$t\left(K_{\underbrace{r,r,\ldots,r}_{k \text{ terms}}}\right) = r^{rk-2} k^{k-2} (k-1)^{(r-1)k}.$$

Finally, if $G = K_{p,p} - pK_2$, then $\overline{G}$ consists of two copies of K_p joined by a matching of size p. Hence, with the node set appropriately labeled, $\mathbf{H}(\overline{G})$ has the block form given below:

$$\mathbf{H}(\overline{G}) = \begin{bmatrix} (p+1)\mathbf{I}_p - \mathbf{J}_p & -\mathbf{I}_p \\ -\mathbf{I}_p & (p+1)\mathbf{I}_p - \mathbf{J}_p \end{bmatrix}.$$

Of course,

$$\lambda\mathbf{I}_{2p} - \mathbf{H}(\overline{G}) = \begin{bmatrix} (\lambda-(p+1))\mathbf{I}_p + \mathbf{J}_p & \mathbf{I}_p \\ \mathbf{I}_p & (\lambda-(p+1))\mathbf{I}_p + \mathbf{J}_p \end{bmatrix}.$$

Since any matrix $\begin{bmatrix} A & B \\ C & D \end{bmatrix}$ with square blocks *A, B, C,* and *D* has determinant equal to det $(AD\text{-}BC)$ if *A* and *C* commute, we may write:

$$\det\left(\lambda\mathbf{I}_{2p} - \mathbf{H}(\overline{G})\right) = \det\left(\left((\lambda-(p+1))\mathbf{I}_p + \mathbf{J}_p\right)^2 - \mathbf{I}_p\right) =$$

$$\det\left(\left((\lambda-(p+1))^2 - 1\right)\mathbf{I}_p - (p+2-2\lambda)\mathbf{J}_p\right).$$

Now consider the general determinant $\det\left(a\mathbf{I}_p - b\mathbf{J}_p\right)$.

If $b = 0$, then $\det\left(a\mathbf{I}_p - b\mathbf{J}_p\right) = a^p$.

If $b \neq 0$, then $\det(a\mathbf{I}_p - b\mathbf{J}_p) = b^p \cdot \det\left(\frac{a}{b}\mathbf{I}_p - \mathbf{J}_p\right)$.

But $\det\left(\lambda\mathbf{I}_p - \mathbf{J}_p\right) = \lambda^{p-1}(\lambda - p)$, so that

$$\det\left(a\mathbf{I}_p - b\mathbf{J}_p\right) = b^p\left(\frac{a}{b}\right)^{p-1}\left(\frac{a}{b} - p\right) = a^{p-1}(a - bp).$$

Thus, whether $b = 0$ or not,

$$\det\left(a\mathbf{I}_p - b\mathbf{J}_p\right) = a^{p-1}(a - bp).$$

Applying this result to the problem at hand, we obtain

$$\det\left(\lambda\mathbf{I}_{2p} - \mathbf{H}(\overline{G})\right) = \left((\lambda-p)^2 - 1\right)^{p-1}\left((\lambda-p)^2 - 1 - (p-2\lambda)p\right)$$

$$= (\lambda - p - 2)^{p-1}(\lambda - p)^{p-1}\left(\lambda^2 - 2\lambda\right).$$

Thus, by Corollary 1.10,

$$t\left(K_{p,p} - pK_2\right) = (p-1)p^{p-2}(p-2)^{p-1}. \qquad \square$$

a) $K_{2,3}$

b) $K_{2,2,2}$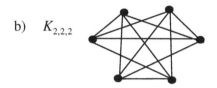

c) $K_{3,3} - 3K_2$

Figure 1.6: *k*-partite graphs.

(c) (i) Observe that the Laplacian matrix of a path is

$$\mathbf{H}\left(P_p\right) = \begin{bmatrix} 1 & -1 & 0 & 0 & \cdots & 0 \\ -1 & 2 & -1 & 0 & \cdots & 0 \\ 0 & -1 & 2 & -1 & \ddots & 0 \\ \vdots & \ddots & \ddots & \ddots & \ddots & 0 \\ 0 & \cdots & 0 & -1 & 2 & -1 \\ 0 & \cdots & 0 & 0 & -1 & 1 \end{bmatrix}.$$

We let $f_p(x)$ denote the determinant of the general $p \times p$ matrix with the following form:

$$\begin{bmatrix} x & -1 & 0 & 0 & \cdots & 0 \\ -1 & x+1 & -1 & 0 & \cdots & 0 \\ 0 & -1 & x+1 & -1 & \ddots & 0 \\ \vdots & \ddots & \ddots & \ddots & \ddots & 0 \\ 0 & \cdots & 0 & -1 & x+1 & -1 \\ 0 & \cdots & 0 & 0 & -1 & x \end{bmatrix}.$$

After subsequent expansion about the first and last rows (respectively), denote by $g_p(y)$ the determinant of the $p \times p$ matrix

$$\begin{bmatrix} y & -1 & 0 & 0 & \cdots & 0 \\ -1 & y & -1 & 0 & \cdots & 0 \\ 0 & -1 & y & -1 & \ddots & 0 \\ \vdots & \ddots & \ddots & \ddots & \ddots & 0 \\ 0 & \cdots & 0 & -1 & y & -1 \\ 0 & \cdots & 0 & 0 & -1 & y \end{bmatrix}.$$

Expanding $g_p(y)$ about the first row yields $g_p(y) = y g_{p-1}(y) - g_{p-2}(y)$.

Of course, this is equivalent to $g_{p-1}(y) - y g_{p-2}(y) + g_{p-3}(y) = 0$.

Setting $y = x + 1$ in this equation yields

$$g_{p-1}(1+x) - (1+x)g_{p-2}(1+x) + g_{p-3}(1+x) = 0.$$

Setting $y = x + 1$ in the first equation and multiplying the result by $(1-x)$ yields

$$(1-x)g_p(1+x) + (x^2 - 1)g_{p-1}(1+x) + (1-x)g_{p-2}(1+x) = 0.$$

Adding the previous two equations and simplifying further gives

$$x^2 g_{p-1}(x+1) - 2x g_{p-2}(x+1) + g_{p-3}(x+1) - (x-1)g_p(x+1) = 0,$$

from which we obtain

$$f_p(x) = x^2 g_{p-2}(x+1) - 2x g_{p-3}(x+1) + g_{p-4}(x+1) =$$
$$(x-1) g_{p-1}(x+1). \qquad (*)$$

Now, if $U_p(z)$ denotes the Chebyshev polynomial of the second kind, which is defined by the recursion:

$$U_{p+2}(z) - 2z U_{p+1}(z) + U_p(z) = 0, \text{ where } U_0(z) = 1 \text{ and } U_1(z) = 2z,$$

then $g_p(y) = U_p\left(\dfrac{y}{2}\right)$. It is known [Seshu, 1961] that

$$U_p(z) = \frac{\left(z + \sqrt{z^2 - 1}\right)^{p+1} - \left(z - \sqrt{z^2 - 1}\right)^{p+1}}{2\sqrt{z^2 - 1}}$$

for $z^2 \neq 1$, $U_p(1) = p+1$, and $U_p(-1) = (-1)^p(p+1)$. Thus, from $(*)$

and $g_p(y) = U_p\left(\dfrac{y}{2}\right)$, we obtain, $f_p(x) = 0$ if $x = 1$, $4p(-1)^p$ if $x = -3$,

and

$$\frac{x-1}{2\sqrt{\left(\dfrac{x+1}{2}\right)^2 - 1}}\left\{\left(\frac{x+1}{2} + \sqrt{\left(\frac{x+1}{2}\right)^2 - 1}\right)^p - \left(\frac{x+1}{2} - \sqrt{\left(\frac{x+1}{2}\right)^2 - 1}\right)^p\right\}$$

if $x \neq 1, -3$. The substitution $x = 1 - \lambda$ yields the characteristic polynomial $\det\left(\lambda \mathbf{I}_p - \mathbf{H}(P_p)\right) = (-1)^p f_p(1 - \lambda) = 0$ if $\lambda = 0$, $4p$ if $\lambda = 4$, and if $\lambda \neq 0, 4$ the determinant is

$$\frac{(-1)^{p+1}\lambda}{2\sqrt{\left(\frac{2-\lambda}{2}\right)^2-1}}\left(\left(\frac{2-\lambda}{2}+\sqrt{\left(\frac{2-\lambda}{2}\right)^2-1}\right)^p-\left(\frac{2-\lambda}{2}-\sqrt{\left(\frac{2-\lambda}{2}\right)^2-1}\right)^p\right).$$

Finally, the result of Corollary 1.10 yields $t(G) = 4^{3-p}\cdot p$ if $n=4$, and

$$\left(n^{n-p-1}\right)\frac{n^{n-p-1}}{2\sqrt{\left(\frac{n-2}{2}\right)^2-1}}\left[\left(\frac{n-2}{2}+\sqrt{\left(\frac{n-2}{2}\right)^2-1}\right)^p-\left(\frac{n-2}{2}-\sqrt{\left(\frac{n-2}{2}\right)^2-1}\right)^p\right]$$

if $n>4$.

(ii) We first consider the Laplacian matrix of the cycle,

$$\mathbf{H}(C_p)=\begin{bmatrix} 2 & -1 & 0 & 0 & \cdots & -1 \\ -1 & 2 & -1 & 0 & \cdots & 0 \\ 0 & -1 & 2 & -1 & \ddots & 0 \\ \vdots & \ddots & \ddots & \ddots & \ddots & 0 \\ 0 & \cdots & 0 & -1 & 2 & -1 \\ -1 & 0 & \cdots & 0 & -1 & 2 \end{bmatrix}.$$

We observe, in a similar fashion to the proof in (c)(i), that this matrix has the form

$$\begin{bmatrix} y & -1 & 0 & 0 & \cdots & -1 \\ -1 & y & -1 & 0 & \cdots & 0 \\ 0 & -1 & y & -1 & \ddots & 0 \\ \vdots & \ddots & \ddots & \ddots & \ddots & 0 \\ 0 & \cdots & 0 & -1 & y & -1 \\ -1 & 0 & \cdots & 0 & -1 & y \end{bmatrix}$$

and its determinant, $h_p(y)$, can be written $h_p(y)=y\,g_{p-1}(y)$ $-2g_{p-2}(y)-2$, where $g_p(y)$ is the determinant of the matrix described in the proof of (c)(i). Similar calculations lead to

$$\det\left(\lambda\mathbf{I}_p-\mathbf{H}(C_p)\right)=(-1)^p h_p(2-\lambda).$$

Corollary 1.10, and direct substitution yield the formula $t(G) =$ $\dfrac{2 - 2(-1)^p}{4}$ if $n = 4$, and $p = 3, 4$, and

$$n^{n-p-2}\left[\left(\frac{n-2}{2} + \sqrt{\left(\frac{n-2}{2}\right)^2 - 1}\right)^p + \left(\frac{n-2}{2} - \sqrt{\left(\frac{n-2}{2}\right)^2 - 1}\right)^p + 2(-1)^{p+1}\right]$$

if $n > 4$.

Figure 1.7: $K_5 - P_3$. The dashed edges indicate the deleted P_3.

Figure 1.8: $K_5 - C_3$. The dashed edges indicate the deleted C_3.

The following example checks the validity of the formulas for the number of spanning trees found in this section. We do not include K_n here, as Cayley's formula is well known.

Example 1.4 We consider the graphs depicted in Figures 0.16c, 0.9, and 1.5-1.8. For each we find $t(G)$ by first applying the formulas found in Corollary 1.8, Proposition 1.9 and Proposition 1.12, and second by computing the $(1,1)$-cofactor of the Laplacian of G.

Example 1.4 a – Product of complete graphs $K_n \times K_m$.

By Corollary 1.8 $t\left(K_n \times K_m\right) = (m+n)^{(n-1)(m-1)} m^{m-2} n^{n-2}$.

For $K_2 \times K_3$ (Figure 0.16c) $n = 2$ and $m = 3$, so

$$t\left(K_2 \times K_3\right) = (3+2)^{(2-1)(3-1)} 3^{3-2} 2^{2-2} = 5^2 3^1 2^0 = 75.$$

One Laplacian matrix for $K_2 \times K_3$ is
$\begin{bmatrix} 3 & -1 & -1 & -1 & 0 & 0 \\ -1 & 3 & -1 & 0 & -1 & 0 \\ -1 & -1 & 3 & 0 & 0 & -1 \\ -1 & 0 & 0 & 3 & -1 & -1 \\ 0 & -1 & 0 & -1 & 3 & -1 \\ 0 & 0 & -1 & -1 & -1 & 3 \end{bmatrix}$ which

has (1,1)-cofactor 75 as well, confirming the result of the formula.

Example 1.4 b – Circulant graph $C_n\left(n_1,...,n_k\right)$.

By Proposition 1.9: $t\left(C_n\left(n_1,...,n_k\right)\right) =$

$$\begin{cases} \dfrac{2^{n-1}}{n} \prod_{j=1}^{n-1}\left[k - \sum_{i=1}^{k}\cos\left(\dfrac{2\pi n_i j}{n}\right)\right] & \text{if } n_k < \dfrac{n}{2} \\ \dfrac{2^{n-1}}{n} \prod_{m=1}^{\frac{n}{2}-1}\left[k - \sum_{i=1}^{k}\cos\left(\dfrac{4\pi n_i m}{n}\right)\right] \prod_{m=1}^{\frac{n}{2}}\left[k-1- \sum_{i=1}^{k}\cos\left(\dfrac{2(2m-1)\pi n_i}{n}\right)\right] & \text{if } n_k = \dfrac{n}{2} \end{cases}.$$

For $C_8(1,2)$ (Figure 0.7) $k = 2$, $n = 8$, $n_1 = 1$, and $n_2 = 2$. Thus, $n_k < \dfrac{n}{2}$

so we use the top formula, and

$$t\left(C_8\left(1,2\right)\right)=\frac{2^{8-1}}{8}\prod_{j=1}^{8-1}\left[2-\sum_{i=1}^{2}\cos\left(\frac{2\pi n_i j}{8}\right)\right]=$$

$$\frac{2^7}{8}\prod_{j=1}^{7}\left[2-\left(\cos\left(\frac{2\pi n_1 j}{8}\right)+\cos\left(\frac{2\pi n_2 j}{8}\right)\right)\right]=$$

$$\frac{2^7}{8}\left[2-\left(\cos\left(\frac{2\pi\left(1\right)\left(1\right)}{8}\right)+\cos\left(\frac{2\pi\left(2\right)\left(1\right)}{8}\right)\right)\right]\times$$

$$\left[2-\left(\cos\left(\frac{2\pi\left(1\right)\left(2\right)}{8}\right)+\cos\left(\frac{2\pi\left(2\right)\left(2\right)}{8}\right)\right)\right]\times$$

$$\left[2-\left(\cos\left(\frac{2\pi\left(1\right)\left(3\right)}{8}\right)+\cos\left(\frac{2\pi\left(2\right)\left(3\right)}{8}\right)\right)\right]\times$$

$$\left[2-\left(\cos\left(\frac{2\pi\left(1\right)\left(4\right)}{8}\right)+\cos\left(\frac{2\pi\left(2\right)\left(4\right)}{8}\right)\right)\right]\times$$

$$\left[2-\left(\cos\left(\frac{2\pi\left(1\right)\left(5\right)}{8}\right)+\cos\left(\frac{2\pi\left(2\right)\left(5\right)}{8}\right)\right)\right]\times$$

$$\left[2-\left(\cos\left(\frac{2\pi\left(1\right)\left(6\right)}{8}\right)+\cos\left(\frac{2\pi\left(2\right)\left(6\right)}{8}\right)\right)\right]\times$$

$$\left[2-\left(\cos\left(\frac{2\pi\left(1\right)\left(7\right)}{8}\right)+\cos\left(\frac{2\pi\left(2\right)\left(7\right)}{8}\right)\right)\right]=3528.$$

One Laplacian matrix for $C_8\left(1,2\right)$ is

$$\begin{bmatrix} 4 & -1 & -1 & 0 & 0 & 0 & -1 & -1 \\ -1 & 4 & -1 & -1 & 0 & 0 & 0 & -1 \\ -1 & -1 & 4 & -1 & -1 & 0 & 0 & 0 \\ 0 & -1 & -1 & 4 & -1 & -1 & 0 & 0 \\ 0 & 0 & -1 & -1 & 4 & -1 & -1 & 0 \\ 0 & 0 & 0 & -1 & -1 & 4 & -1 & -1 \\ -1 & 0 & 0 & 0 & -1 & -1 & 4 & -1 \\ -1 & -1 & 0 & 0 & 0 & -1 & -1 & 4 \end{bmatrix}$$

which has (1,1)-cofactor 3528 as well, confirming the result of the formula.

Example 1.4 c – K_n minus a matching.

By Proposition 1.12 a): $t\left(K_n - kK_2\right) = n^{n-2}\left(1 - \dfrac{2}{n}\right)^k$.

For $K_6 - 2K_2$ (Figure 1.5), $n = 6$ and $k = 2$, so

$$t\left(K_6 - 2K_2\right) = 6^{6-2}\left(1 - \frac{2}{6}\right)^2 = 6^4\left(\frac{4}{9}\right) = 576.$$

One Laplacian matrix for $K_6 - 2K_2$ is

$$\begin{bmatrix} 4 & -1 & -1 & -1 & 0 & -1 \\ -1 & 4 & -1 & 0 & -1 & -1 \\ -1 & -1 & 5 & -1 & -1 & -1 \\ -1 & 0 & -1 & 4 & -1 & -1 \\ 0 & -1 & -1 & -1 & 4 & -1 \\ -1 & -1 & -1 & -1 & -1 & 5 \end{bmatrix}$$

which has (1,1)-cofactor 576 as well, confirming the result of the formula.

Example 1.4 d – Complete Bipartite $K_{p,q}$.

By Proposition 1.12 b i): $t\left(K_{p,q}\right) = q^{p-1}p^{q-1}$.

For $K_{2,3}$ (Figure 1.6 a) $p = 2$ and $q = 3$, so $t\left(K_{2,3}\right) = 3^{2-1}2^{3-1} = 12.$

One Laplacian matrix for $K_{2,3}$ is $\begin{bmatrix} 3 & 0 & -1 & -1 & -1 \\ 0 & 3 & -1 & -1 & -1 \\ -1 & -1 & 2 & 0 & 0 \\ -1 & -1 & 0 & 2 & 0 \\ -1 & -1 & 0 & 0 & 2 \end{bmatrix}$ which has

(1,1)-cofactor 12 as well, confirming the result of the formula.

Example 1.4 e – Regular complete k-partite $K_{r,r,\ldots r}$.

By Proposition 1.12 b ii): $t\left(\underbrace{K_{r,r,\ldots r}}_{k \text{ terms}} \right) = r^{rk-2}k^{k-2}\left(k-1\right)^{(r-1)k}$.

For $K_{2,2,2}$ (Figure 1.6 b) $k = 3$ and $r = 2$, so

$$t\left(K_{2,2,2}\right) = 2^{2(3)-2}3^{3-2}\left(3-1\right)^{(2-1)3} = 2^4 \cdot 3 \cdot 2^3 = 384.$$

One Laplacian matrix for $K_{2,2,2}$ is $\begin{bmatrix} 4 & 0 & -1 & -1 & -1 & -1 \\ 0 & 4 & -1 & -1 & -1 & -1 \\ -1 & -1 & 4 & 0 & -1 & -1 \\ -1 & -1 & 0 & 4 & -1 & -1 \\ -1 & -1 & -1 & -1 & 4 & 0 \\ -1 & -1 & -1 & -1 & 0 & 4 \end{bmatrix}$ which has (1,1) cofactor 384 as well, confirming the result of the formula.

Example 1.4 f – Complete bipartite with equal part sizes minus a full matching $K_{p,p} - pK_2$.

By Proposition 1.12 b iii): $t\left(K_{p,p} - pK_2\right) = (p-1)\, p^{p-2}\, (p-2)^{p-1}$.

For $K_{3,3} - 3K_2$ (Figure 1.6 c) $p = 2$ so

$$t\left(K_{3,3} - 3K_2\right) = (3-1)\cdot 3^{3-2}\cdot (3-2)^{3-1} = 6.$$

One Laplacian matrix for $K_{3,3} - 3K_2$ is

$$\begin{bmatrix} 2 & 0 & 0 & 0 & -1 & -1 \\ 0 & 2 & 0 & -1 & 0 & -1 \\ 0 & 0 & 2 & -1 & -1 & 0 \\ 0 & -1 & -1 & 2 & 0 & 0 \\ -1 & 0 & -1 & 0 & 2 & 0 \\ -1 & -1 & 0 & 0 & 0 & 2 \end{bmatrix}$$

which has (1,1)-cofactor 6 as well, confirming the result of the formula.

Example 1.4 g – Complete graph of order n minus a path of order p $(n > p)$ $K_n - P_p$.

By Proposition 1.12 c i):

$$t\left(K_n - P_p\right) = \begin{cases} \dfrac{n^{n-p-2}}{2\sqrt{\left(\dfrac{n-2}{2}\right)^2 - 1}}\left[\left(\dfrac{n-2}{2} + \sqrt{\left(\dfrac{n-2}{2}\right)^2 - 1}\right)^{p+1} - \left(\dfrac{n-2}{2} - \sqrt{\left(\dfrac{n-2}{2}\right)^2 - 1}\right)^{p+1}\right] & \text{if } n > 4 \\ \\ 4^{3-p}\cdot p & \text{if } n = 4 \end{cases}.$$

For $K_5 - P_3$ (Figure 1.7) $n = 5$ and $p = 3$ thus the top formula is used, so

$t\left(K_5 - P_3\right) =$

$$\frac{5^{5-3-2}}{2\sqrt{\left(\frac{5-2}{2}\right)^2 - 1}}\left[\left(\frac{5-2}{2} + \sqrt{\left(\frac{5-2}{2}\right)^2 - 1}\right)^{3+1} - \left(\frac{5-2}{2} - \sqrt{\left(\frac{5-2}{2}\right)^2 - 1}\right)^{3+1}\right] = 21.$$

One Laplacian matrix for $K_5 - P_3$ is $\begin{bmatrix} 3 & -1 & -1 & 0 & -1 \\ -1 & 3 & 0 & -1 & -1 \\ -1 & 0 & 2 & 0 & -1 \\ 0 & -1 & 0 & 2 & -1 \\ -1 & -1 & -1 & -1 & 4 \end{bmatrix}$ which has

(1,1)-cofactor 21 as well, confirming the result of the formula.

Example 1.4 h – Complete graph of order n minus a cycle of order p $(n > p)$ $K_n - C_p$.

By Proposition 1.12 c ii): $t\left(K_n - C_p\right) =$

$$\begin{cases} n^{n-p-2}\left[\left(\frac{n-2}{2} + \sqrt{\left(\frac{n-2}{2}\right)^2 - 1}\right)^p + \left(\frac{n-2}{2} - \sqrt{\left(\frac{n-2}{2}\right)^2 - 1}\right)^p + 2(-1)^{p+1}\right] & \text{if } n > 4 \\ \frac{2 - 2(-1)^p}{4} & \text{if } n = 4 \text{ and } p = 3,4 \end{cases}$$

For $K_5 - C_3$ (Figure 1.8) $n = 5$ and $p = 3$, thus the top formula is used, so $t\left(K_5 - C_3\right) =$

$$5^{5-3-2}\left[\left(\frac{5-2}{2} + \sqrt{\left(\frac{5-2}{2}\right)^2 - 1}\right)^3 - \left(\frac{5-2}{2} - \sqrt{\left(\frac{5-2}{2}\right)^2 - 1}\right)^3\right] = 47.$$

One Laplacian matrix for $K_5 - C_3$ is $\begin{bmatrix} 2 & -1 & 0 & 0 & -1 \\ -1 & 4 & -1 & -1 & -1 \\ 0 & -1 & 2 & 0 & -1 \\ 0 & -1 & 0 & 2 & -1 \\ -1 & -1 & -1 & -1 & 4 \end{bmatrix}$ which has

(1,1)-cofactor 47 as well, confirming the result of the formula.

1.4   Temperley's B-Matrix

By Proposition 1.1 and Theorem 1.3, the Laplacian matrix **H** is a singular matrix whose cofactors are all equal. Sometimes it is useful to work with a non-singular matrix derived from **H**. Specifically, the addition of **J**, where **J** is a square matrix with all entries equal to 1, to **H** will yield such a non-singular matrix, called Temperley's **B-Matrix**, or just **B** [Temperley, 1964].

In classes of graphs with many edges, **B**'s eigenvalues are easier to compute, as there are zero elements wherever a "−1" appears in **H**. The eigenvalues of **B** and **H** are the same, save a zero eigenvalue of **H** and an extra eigenvalue of n in **B**. This fact is an immediate corollary of the next matrix-theoretic result:

Theorem 1.13 *If* **L** *is a matrix whose rows or columns sum to zero, then the characteristic polynomials of* **L** *and* **L** + g**J**   *(g an integer), respectively,* $P_{\mathbf{L}}(\lambda)$ *and* $P_{\mathbf{L}+g\mathbf{J}}(\lambda)$ *are related as follows:*

$$P_{\mathbf{L}}(\lambda)(\lambda - gn) = P_{\mathbf{L}+g\mathbf{J}}(\lambda)\lambda.$$

Proof Without loss of generality, assume all rows of **L** sum to zero.

$$P_{\mathbf{L}}(\lambda) = \det(\lambda\mathbf{I} - \mathbf{L}) = \det \begin{bmatrix} \lambda - l_{11} & -l_{12} & -l_{13} & \cdots & -l_{1n} \\ -l_{21} & \lambda - l_{22} & -l_{23} & \cdots & -l_{2n} \\ \vdots & \vdots & \vdots & \ddots & \vdots \\ -l_{n1} & -l_{n2} & -l_{n3} & \cdots & \lambda - l_{nn} \end{bmatrix}.$$

Addition of all the columns to the first yields:

$$P_{\mathbf{L}}(\lambda) = \det \begin{bmatrix} \lambda & -l_{12} & -l_{13} & \cdots & -l_{1n} \\ \lambda & \lambda - l_{22} & -l_{23} & \cdots & -l_{2n} \\ \vdots & \vdots & \vdots & \ddots & \vdots \\ \lambda & -l_{n2} & -l_{n3} & \cdots & \lambda - l_{nn} \end{bmatrix} =$$

$$\lambda \det \begin{bmatrix} 1 & -l_{12} & -l_{13} & \cdots & -l_{1n} \\ 1 & \lambda - l_{22} & -l_{23} & \cdots & -l_{2n} \\ \vdots & \vdots & \vdots & \ddots & \vdots \\ 1 & -l_{n2} & -l_{n3} & \cdots & \lambda - l_{nn} \end{bmatrix} = \lambda \sum_{i=1}^{n} C_{i1}(\lambda\mathbf{I} - \mathbf{L}).$$

Similarly, consider $P_{\mathbf{L}+g\mathbf{J}}(\lambda) =$

$$\det \begin{bmatrix} \lambda - l_{11} - g & -l_{12} - g & -l_{13} - g & \cdots & -l_{1n} - g \\ -l_{21} - g & \lambda - l_{22} - g & -l_{23} - g & \cdots & -l_{2n} - g \\ \vdots & \vdots & \vdots & \ddots & \vdots \\ -l_{n1} - g & -l_{n2} - g & -l_{n3} - g & \cdots & \lambda - l_{nn} - g \end{bmatrix}.$$

Addition of all columns to the first yields:

$$\det \begin{bmatrix} \lambda - gn & -l_{12} - g & -l_{13} - g & \cdots & -l_{1n} - g \\ \lambda - gn & \lambda - l_{22} - g & -l_{23} - g & \cdots & -l_{2n} - g \\ \vdots & \vdots & \vdots & \ddots & \vdots \\ \lambda - gn & -l_{n2} - g & -l_{n3} - g & \cdots & \lambda - l_{nn} - g \end{bmatrix}.$$

Factoring out the first column leaves all ones there, and the subsequent addition of g times this column to all the others yields:

$$(\lambda - gn)\det\begin{bmatrix} 1 & -l_{12} & -l_{13} & \cdots & -l_{1n} \\ 1 & \lambda - l_{22} & -l_{23} & \cdots & -l_{2n} \\ \vdots & \vdots & \vdots & \ddots & \vdots \\ 1 & -l_{n2} & -l_{n3} & \cdots & \lambda - l_{nn} \end{bmatrix}.$$

Therefore, $\displaystyle\sum_{i=1}^{n} C_{i1}\left(\lambda \mathbf{I} - \mathbf{L}\right) = \frac{1}{\lambda}P_{\mathbf{L}}\left(\lambda\right) = \frac{1}{\lambda - gn}P_{\mathbf{L}+g\mathbf{J}}\left(\lambda\right).$ □

Corollary 1.14 follows immediately:

Corollary 1.14 *The characteristic polynomials for a Laplacian matrix* **H** *and its corresponding* **B** *-matrix differ by a single factor in each, i.e.*

$$\frac{1}{\lambda}P_{\mathbf{H}}\left(\lambda\right) = \frac{1}{\lambda - n}P_{\mathbf{B}}\left(\mathbf{B},\lambda\right).$$

Proof In Theorem 1.13 setting $\mathbf{L} = \mathbf{H}$ and $g = 1$, then

$$P_{\mathbf{H}}\left(\lambda\right)\left(\lambda - n\right) = P_{\mathbf{L}}\left(\lambda\right)\left(\lambda - n\right) = P_{\mathbf{L}+\mathbf{J}}\left(\lambda\right)\lambda = P_{\mathbf{B}}\left(\lambda\right)\lambda \qquad □$$

Theorem 1.15 *Suppose the complement of the graph G consists of k node disjoint subgraphs $\mathcal{C}_1, \mathcal{C}_2, ..., \mathcal{C}_k$ so that G is the join of $\bar{\mathcal{C}}_1, \bar{\mathcal{C}}_2, ..., \bar{\mathcal{C}}_k$, the complements on $n_1, n_2, ..., n_k$ nodes, respectively. If $G_i = \bar{\mathcal{C}}_i + K_{n-n_i}$ for each i, then $t(G) = \dfrac{1}{n^{(k-1)n-2(k-1)}}\displaystyle\prod_{i=1}^{k} t(G_i)$. In particular, if $k = 2$, $t(G) = \dfrac{1}{n^{n-2}}t(G_1)t(G_2)$.*

Proof   Realize that $\mathbf{H}(G) = \begin{bmatrix} \mathbf{H}(\overline{C}_1) & -1 & \cdots & -1 \\ -1 & \mathbf{H}(\overline{C}_2) & -1 & \vdots \\ \vdots & -1 & \ddots & -1 \\ -1 & \cdots & -1 & \mathbf{H}(\overline{C}_k) \end{bmatrix}$ so that

$$\mathbf{B}(G) = \begin{bmatrix} \mathbf{B}(\overline{C}_1) & 0 & \cdots & 0 \\ 0 & \mathbf{B}(\overline{C}_2) & 0 & \vdots \\ \vdots & 0 & \ddots & 0 \\ 0 & \cdots & 0 & \mathbf{B}(\overline{C}_k) \end{bmatrix}\text{ and }$$

$$t(G) = \frac{1}{n^2}\det\left(\mathbf{B}(G)\right) = \frac{1}{n^2}\prod_{i=1}^{k}\det\left(\mathbf{B}(\overline{C}_i)\right).$$

Now, $\mathbf{B}(G_i) = \begin{bmatrix} \mathbf{B}(\overline{C}_i) & 0 \\ 0 & n\mathbf{I}_{n-n_i} \end{bmatrix}$, so $\det\left(\mathbf{B}(G_i)\right) = n^{n-n_i}\det\left(\mathbf{B}(\overline{C}_i)\right)$ and

$t(G_i) = \frac{1}{n^2}\det\left(\mathbf{B}(G_i)\right) = n^{n-n_i-2}\det\left(\mathbf{B}(\overline{C}_i)\right).$ Thus, $\prod_{i=1}^{k}t(G_i) =$

$\prod_{i=1}^{k}n^{n-n_i-2}\det\left(\mathbf{B}(\overline{C}_i)\right) = n^{kn-n-2k}\prod_{i=1}^{k}\det\left(\mathbf{B}(\overline{C}_i)\right) = n^{(k-1)n-2k}n^2 t(G).$

Therefore, $t(G) = \dfrac{1}{n^{(k-1)n-2(k-1)}}\prod_{i=1}^{k}t(G_i).$ □

The following general theorem suggests a relationship between the number of spanning trees and the cofactors of $\mathbf{B}$.

Theorem 1.16 *Given any equicofactor matrix* $\mathbf{L}$, *where C denotes the common value of all cofactors of H, and $C \neq 0$, then*

$$\det\left(\mathbf{L} + g\mathbf{J}\right) = gn^2 C(\mathbf{L}).$$

Proof By Remark 1.2 (following the proof of Theorem 1.3), $\mathbf{L}$ has row and column sum zero. Thus, the addition of all columns of $\mathbf{L} + g\mathbf{J}$ to the first yields:

$$\det \begin{bmatrix} gn & l_{12} + g & l_{13} + g & \cdots & l_{1n} + g \\ gn & l_{22} + g & l_{23} + g & \cdots & l_{2n} + g \\ \vdots & \vdots & \vdots & \ddots & \vdots \\ gn & l_{n2} + g & l_{n3} + g & \cdots & l_{nn} + g \end{bmatrix} =$$

$$gn \det \begin{bmatrix} 1 & l_{12} + g & l_{13} + g & \cdots & l_{1n} + g \\ 1 & l_{22} + g & l_{23} + g & \cdots & l_{2n} + g \\ \vdots & \vdots & \vdots & \ddots & \vdots \\ 1 & l_{n2} + g & l_{n3} + g & \cdots & l_{nn} + g \end{bmatrix}.$$

Subtraction of g times column 1 from all the other columns yields

$$gn \det \begin{bmatrix} 1 & l_{12} & l_{13} & \cdots & l_{1n} \\ 1 & l_{22} & l_{23} & \cdots & l_{2n} \\ \vdots & \vdots & \vdots & \ddots & \vdots \\ 1 & l_{n2} & l_{n3} & \cdots & l_{nn} \end{bmatrix} = gn^2 C(\mathbf{L}) \quad \text{upon expansion about column}$$

one.   □

Theorem 1.16 leads directly to the following result:

Corollary 1.17 *The number of spanning trees for a graph can be determined from its B-matrix by the formula*

$$\det(\mathbf{B}) = n^2 C(\mathbf{H}) = n^2 (t(G)).$$

1.5 Multigraphs

Most of the classical results regarding the eigenvalues of the Laplacian matrix are only valid for graphs because they depend on the existence of the complement of a graph, which is formed by removing the edges of the graph in question from the complete graph. It is easy to see that the Laplacian matrix of the complement of a graph is given by $\bar{\mathbf{H}} = n\mathbf{I} - \mathbf{J} - \mathbf{H}$. Henceforth, we shall refer to $\bar{\mathbf{H}}$ given as above, as the *algebraic complement* of $\mathbf{H}$ (or G) even in the event that G is a multigraph. In the case of a multigraph, $\bar{\mathbf{H}}$ has no graph theoretical interpretation.

Some of the useful results one obtains concerning $\mathbf{H}$ and $\mathbf{B}$ in the graph case are lost when multigraphs are considered. For example, since the Laplacian matrix of a multigraph is non-negative definite, symmetric and real, the eigenvalues are non-negative. Hence, in the event that G is a graph, so that $\bar{\mathbf{H}}$ is the Laplacian matrix of the complement $\bar{G}$, we claim that the maximum eigenvalue of $\mathbf{H}$ and $\mathbf{B}$ is n. Indeed, the eigenvalues of $\mathbf{B} = n\mathbf{I} - \mathbf{H}$ are given by $n - \lambda(\bar{\mathbf{H}})$ where $\lambda(\bar{\mathbf{H}})$ runs through the eigenvalues of $\bar{\mathbf{H}}$.

As the minimum eigenvalue of $\bar{\mathbf{H}}$ is 0, the maximum eigenvalue of $\mathbf{B}$ (and thus of $\mathbf{H}$ as well) is n. This fact does not generalize to multigraphs, as demonstrated by the multigraph given in Figure 1.9.

Figure 1.9: A multigraph in the class $\Omega(5,8)$.

Its Laplacian matrix is $\mathbf{H} = \begin{bmatrix} 3 & -1 & -1 & -1 & 0 \\ -1 & 3 & -1 & -1 & 0 \\ -1 & -1 & 3 & 0 & -1 \\ -1 & -1 & 0 & 4 & -2 \\ 0 & 0 & -1 & -2 & 3 \end{bmatrix}$ which has

maximum eigenvalue approximately equal to 6.214 > 5. Another well-known result for graphs states that $\lambda_n = n$ if and only if the complement is disconnected (see Proposition 1.6), and the multiplicity of n as an eigenvalue of $\mathbf{H}$ is the number of components of $\overline{G}$ [Kelmans 1967]. This result has no interpretation in the multigraph setting.

The bounds of Fiedler on the second eigenvalue, namely, $\lambda_2 \leq \kappa$ (node connectivity) and $\lambda_2 \leq \delta$ (minimum node degree), respectively, (see Proposition 1.4 (c) and Remark 1.4) are invalid in the multigraph case. A counterexample is given by the multigraph in Figure 1.10. For this graph, $\kappa = 2$ and $\delta = 4$, and its Laplacian matrix has $\lambda_2 = 6$.

**Figure 1.10: A multigraph that serves as a counterexample to the bounds
of Fiedler and Kelmans.**

Another eigenvalue bound which does not apply to multigraphs states that $\lambda_k \geq \chi$ where K_χ is the largest complete subgraph and $n - \chi - 2 \leq k \leq n$. Consider the multigraph given in Figure 1.11.

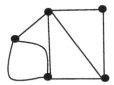

**Figure 1.11: A multigraph that serves as a counterexample to a
graph eigenvalue bound.**

For this graph, $\chi = 3$ and $n - \chi - 2 = 5 - 3 - 2 \le 2 \le 5$ but we have
$\lambda_2 \approx 1.6571 < \chi$.

1.6   Eigenvalue Bounds for Multigraphs

We proceed to derive some bounds on the eigenvalues of the Laplacian
matrix of a multigraph, which generalize known results for graphs.

In multigraphs where the largest multiplicity of any edge is g, it is
sometimes useful to form a "new" B-matrix, $\hat{\mathbf{B}}$ by adding $g\,\mathbf{J}$ instead of
$\mathbf{J}$ to $\mathbf{H}$. An immediate consequence of Theorem 1.13 is the following
corollary:

Corollary 1.18 *Let M be a multigraph such that the largest multiplicity
of any edge is g. Given* $\mathbf{H}$ *and* $\hat{\mathbf{B}} = \mathbf{H} + g\mathbf{J}$, *then*

$$\frac{1}{\lambda} P_{\mathbf{H}}(\lambda) = \frac{1}{\lambda - gn} P_{\hat{\mathbf{B}}}(\lambda),$$

and   the   number   of   spanning   trees   is   given   by

$$\frac{1}{gn^2}\prod_{i=1}^{n}\lambda_i\left(\hat{\mathbf{B}}\right), \quad \text{where } \lambda_1\left(\hat{\mathbf{B}}\right),...,\lambda_n\left(\hat{\mathbf{B}}\right) \text{ are the eigenvalues of } \hat{\mathbf{B}}.$$

The second eigenvalue for a graph is known to be bounded by $2\delta-(n-2)\le\lambda_2\le\kappa$ (Proposition 1.6 (c)). For multigraphs, we shall prove that $2\delta-g(n-2)\le\lambda_2\le g\kappa$. The proof of $\lambda_2\le g\kappa$ is deferred until later in the discussion.

Proposition 1.19 *If* $\mathbf{H}(M)$ *is the Laplacian matrix of multigraph M with eigenvalues* $0=\lambda_1(M)\le\lambda_2(M)\le\cdots\le\lambda_n(M)$, *then*

$$\lambda_2(M)\ge2\delta-g(n-2),$$

where the largest multiplicity of any edge is g. Furthermore, this bound is best possible.

Proof   Consider $\hat{b}_{ii}-\sum_{j\ne i}\left|\hat{b}_{ij}\right|=h_{ii}+g-\sum_{j\ne i}\left|h_{ij}+g\right|$.   Note   that

$\sum_{j\ne i}\left|h_{ij}+g\right|=g(n-1)-d_i$, since $\left|h_{ij}+g\right|=g-\alpha_{ij}$, where α_{ij} is the

number of parallel edges between nodes i and j. Since $\sum_{j\ne i}\alpha_{ij}=d_i$

where $h_{ii}=d_i$, we have, by the bound of Gersgorin (Corollary 0.8),

$\lambda_0\left(\hat{\mathbf{B}}\right)\ge\hat{b}_{ii}-\sum_{j\ne i}\left|\hat{b}_{ij}\right|$. Thus,

$$\lambda_1(M)\ge\min_i\left\{d_i+g-\left(g(n-1)-d_i\right)\right\}=$$
$$\min_i\left\{2d_i-g(n-2)\right\}\ge2\delta-g(n-2).$$

Now, to see that the bound is best possible consider $g\left[K_k + \left(K_a \cup K_a\right)\right]$ depicted in Figure 1.12, where gG is the graph obtained from G by replicating each edge g times. Here,

$$2\delta - g(n-2) = 2g(a-1+k) - g(k+2a-2) = gk = g\kappa. \qquad \square$$

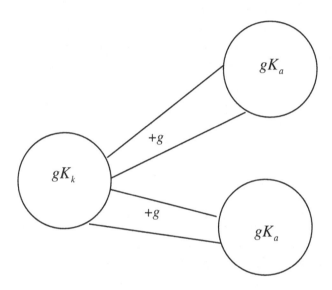

Figure 1.12: The graph that realizes the bounds. Note that there are g edges between each node in gK_k and each node in both gK_a.

In preparation for our next result, we determine the characteristic polynomial of the Laplacian of a multistar. A *multistar* is a multigraph having a star as its underlying graph. In a graph, $\Delta(G)$ is the maximum degree, which is also the number of leaf nodes in the star of largest order that is a subgraph of the graph.

Proposition 1.20 *Let M be a multistar and let m_i be the number of edges in M between the center node and i^{th} leaf node of the underlying*

star. The characteristic polynomial for the Laplacian matrix **H** of M is
given by

$$P_{\mathbf{H}(M)}(\lambda) = \lambda\left(\prod_{i=1}^{\Delta}(\lambda - m_i) - \sum_{i=1}^{\Delta} m_i \prod_{i \neq j}^{\Delta}(\lambda - m_j)\right),$$

where Δ represents the maximum degree of the underlying star.

Proof A multistar as described above has Laplacian matrix

$$\mathbf{H} = \begin{bmatrix} \left(\sum\limits_{i=1}^{\Delta} m_i\right) & -m_1 & \cdot & \cdot & -m_\Delta \\ -m_1 & m_1 & 0 & \cdot & 0 \\ -m_2 & 0 & m_2 & 0 & \cdot \\ \cdot & \cdot & \cdot & \cdot & \cdot \\ -m_\Delta & 0 & \cdot & 0 & m_\Delta \end{bmatrix}.$$

This matrix has characteristic polynomial

$$P_{\mathbf{H}(M)}(\lambda) = \prod_{i=1}^{\Delta}(\lambda - m_i)\left[\lambda - \sum_{i=1}^{\Delta} m_i - \sum_{i=1}^{\Delta}\frac{(m_i)^2}{(\lambda - m_i)}\right]$$

because

$$\det\begin{bmatrix} a & x_1 & x_2 & \cdot & x_n \\ x_1 & y_1 & 0 & \cdot & 0 \\ x_2 & 0 & y_2 & 0 & \cdot \\ \cdot & \cdot & \cdot & \cdot & \cdot \\ x_n & 0 & \cdot & 0 & y_n \end{bmatrix} = a\prod_{i=1}^{n} y_i - \sum_{i=1}^{n}(x_i)^2\prod_{i \neq j}^{n} y_j.$$

Now, $P_{\mathbf{H}(M)}(\lambda)$ simplifies to:

$$\prod_{i=1}^{\Delta}(\lambda - m_i)\left(\lambda - \lambda\sum_{i=1}^{\Delta}\frac{m_i}{(\lambda - m_i)}\right) = \lambda\left(\prod_{i=1}^{\Delta}(\lambda - m_i) - \sum_{i=1}^{\Delta}m_i\prod_{i\neq j}^{\Delta}(\lambda - m_j)\right).$$

$\square$

A multistar with underlying graph $K_{1,q}$ whose q leaf nodes all have common degree, say f, is denoted $fK_{1,q}$. Such a graph has eigenvalues 0, f with multiplicity $q-1$, and $(1+q)f$. Given a multigraph M with underlying graph G, let $u_1, u_2, \ldots, u_r$ be the nodes of G of maximum degree Δ. For each u_i consider the multistar centered at u_i with all leaf nodes having degree f_i equal to the minimum multiplicity of all the edges incident at u_i. We denote this multistar by $f_i K_{1,\Delta}$.

Theorem 1.21 *Let M be a multigraph and Δ be the maximum degree of the underlying graph. Furthermore let f be the largest value such that $fK_{1,\Delta}$ is a submultigraph of M. The lower bound on the largest eigenvalue of the Laplacian matrix of M is $\lambda_n \geq (\Delta+1)f$.*

Proof Express the Laplacian matrix of a multigraph as the sum of the Laplacian matrix of the spanning subgraph formed by the edges of the $fK_{1,\Delta}$ contained in that graph and the Laplacian matrix of the spanning subgraph formed by the remaining edges. The $fK_{1,\Delta}$ Laplacian matrix has eigenvalues as described above, i.e. 0, f with multiplicity $\Delta - 1$, and $(1+\Delta)f$. Referring to Corollary 0.6, let **B** represent the Laplacian matrix for the multstar and **A** the Laplacian matrix for the rest of the graph. Then $\lambda_n(M) \geq (1+\Delta)f$.   □

Example 1.5   Consider the multigraph given in Figure 1.13; its underlying graph has $\Delta = 3$ and contains $3K_{1,3}$ as a submultigraph. The largest eigenvalue is $13.424 > (1+\Delta)f = (1+3)3 = 12$.

Figure 1.13: A multigraph containing $3K_{1,3}$.

Remark 1.6   This subsumes Kelmans' result [Kelmans, 1974], which states that the largest eigenvalue of the Laplacian of a graph is $\Delta + 1$, because in a graph, $f = 1$.

1.7   Multigraph Complements

Consider a multigraph $M = (V,m)$ of order n having Laplacian matrix **H**, and g the maximum edge multiplicity of M. Consider gK_n on V, and define $\overline{M}^{(g)}$ to be the multigraph obtained by removing all the edges of M from gK_n. The following set of results extend eigenvalue results to multigraphs, where g is as stated and f the minimum edge multiplicity.

Theorem 1.22   *If* $0 = \lambda_1 \leq \lambda_2 \leq ... \leq \lambda_n$ *are the eigenvalues of the Laplacian matrix of some multigraph M with maximum edge multiplicity g, then* $\lambda_i\left(\bar{M}^{(g)}\right) = ng - \lambda_{n-i+2}$ *for* $2 \leq n$ *are the eigenvalues of* $\mathbf{H}\left(\bar{M}^{(g)}\right) = g(n\mathbf{I} - \mathbf{J}) - \mathbf{H}(M)$. *Furthermore,* $\lambda_n = ng$ *if and only if* $\bar{M}^{(g)}$ *is disconnected.*

Proof   For ease of notation, let $\mathbf{H} = \mathbf{H}(M)$ and $\bar{\mathbf{H}} = \mathbf{H}\left(\bar{M}^{(g)}\right)$. Let λ be a nonzero eigenvalue of $\mathbf{H}$, and $\mathbf{x}$ an associated eigenvector, i.e., $\mathbf{Hx} = \lambda\mathbf{x}$. Then $\mathbf{1}^T \mathbf{H} = \mathbf{0}$ implies $\mathbf{1}^T \mathbf{x} = 0$. Thus, it follows from $\mathbf{Hx} + \bar{\mathbf{H}}\mathbf{x} = g(n\mathbf{I} - \mathbf{J})\mathbf{x}$ that $\bar{\mathbf{H}}\mathbf{x} = (gn - \lambda)\mathbf{x}$. Likewise, $\mathbf{H1} + \bar{\mathbf{H}}\mathbf{1} = 0 + 0 = 0$. Hence, if $\mathbf{1}, \mathbf{x}_2, \mathbf{x}_3, ..., \mathbf{x_n}$ is a basis for the eigenvectors for $\mathbf{H}$, then they also constitute a basis for the eigenvectors of $\bar{\mathbf{H}}$. Thus, the eigenvalues of $\bar{\mathbf{H}}$ are nonnegative and equal $ng - \lambda_n \leq ng - \lambda_{n-1} \leq \cdots \leq ng - \lambda_1$. It is known that a multigraph, in particular $\bar{M}^{(g)}$ is disconnected if and only if $rank(\bar{\mathbf{H}}) \leq n - 2$. Therefore, $ng - \lambda_n(M) = 0$ if and only if $\bar{M}$ is disconnected.   □

Theorem 1.23   *If M_k is obtained by removing any k nodes from a multigraph M with maximum edge multiplicity g, then* $\lambda_2(M_k) \geq \lambda_2(M) - gk$.

Proof   Without loss of generality, assume M is connected, or else $\lambda_2 = 0$, and the inequality follows trivially. The proof proceeds by induction on k, the number of removed nodes.

For the base case, $k = 1$, let M_1 denote the multigraph obtained from M by removing a single node, and M' be obtained from M_1 by adding a node joined to each of the remaining nodes $v_2, v_3, \ldots, v_n$ by g edges. The Laplacian matrix of M' has $(1,1)$-entry $g(n-1)$, and all the other entries in the first row and column are $-g$. Furthermore,

$$\mathbf{H}_{11}(M') = \mathbf{H}_{11}(M) + g\mathbf{I} = \mathbf{H}(M_1) + g\mathbf{I}.$$

Let $\mathbf{x}$ be the eigenvector of $\bar{\mathbf{H}}(M_1)$ corresponding to λ_2, so that $\mathbf{H}\mathbf{x} = \lambda_2\mathbf{x}$. Now,

$$\mathbf{H}(M')\begin{bmatrix} 0 \\ \mathbf{x} \end{bmatrix} = \begin{bmatrix} -g\mathbf{1}^T\mathbf{x} \\ (\lambda_2(M_1) + g)\mathbf{x} \end{bmatrix} = (\lambda_2(M_1) + g)\begin{bmatrix} 0 \\ \mathbf{x} \end{bmatrix}$$

as $\lambda_2 \geq 0$ implies $\mathbf{1}^T\mathbf{x} = 0$. Hence, $\lambda_2(M_1) + g$ is an eigenvalue of $\mathbf{H}(M')$ and $\lambda_2(M_1) + g \geq \lambda_2(M')$, by Corollary 0.6.

Next, assume that $\lambda_2(M_{k-1}) + g(k-1) \geq \lambda_2(M)$, where M_{k-1} is the multigraph obtained by the removal of $k-1$ nodes from M. Consider a multigraph M_k obtained from M by the removal of any k nodes from M, and let M_{k-1} denote the multigraph obtained by removing any $k-1$ of those nodes from M. It remains to show that $\lambda_2(M_k) + g \geq \lambda_2(M_{k-1})$. But this follows exactly as in the base case in the event that M_{k-1} is connected. If M_{k-1} is not connected, the inequality follows trivially. $\square$

Note: the preceding proof is a minor variation on Fiedler's proof, which applied only to $g = 1$. [Fieldler, 1973]

Theorem 1.23 has two immediate corollaries:

Corollary 1.24 *If M is a connected multigraph with maximum edge multiplicity g and node connectivity κ, then $\lambda_2(M) \leq g\kappa$.*

Proof Let S be a set of nodes of order κ such that $M_\kappa = M - S$. Since M_κ is disconnected $\lambda_2(M_\kappa) = 0$. By Theorem 1.23 $\lambda_2(M_\kappa) \geq \lambda_2(M) - g\kappa$ it follows that that $\lambda_2(M) \leq g\kappa$. □

Remark 1.7 Combining the results of Proposition 1.19 and Theorem 1.23, we have $2\delta - g(n-2) \leq \lambda_2(M) \leq g\Delta$, where Δ is the maximum degree of the underlying graph. Examination of the proof of the fact that $2\delta - g(n-2) \leq \lambda_2(M)$ is best possible shows that there is a multigraph for $g\Delta = 2\delta - g(n-2)$. Hence, $\lambda_2(M) \leq g\Delta$ is also best possible.

Corollary 1.25 *The largest eigenvalue of a multigraph M is bounded from below: $\lambda_n(M) \geq g\left(n - \Delta\left(\bar{M}^{(g)}\right)\right)$, where Δ is the maximum degree of the underlying graph.*

Proof $\lambda_n(M) = ng - \lambda_2\left(\bar{M}^{(g)}\right) \geq ng - g\Delta\left(\bar{M}^{(g)}\right) = g\left(n - \Delta\left(\bar{M}^{(g)}\right)\right)$. □

There are some results for the Laplacian matrix that rely on some techniques of matrix theory. The Laplacian matrix, besides being real and obviously symmetric, is positive semi-definite, which means that

$\mathbf{x}^T \mathbf{H} \mathbf{x} \geq 0$ for all n-dimensional column vectors $\mathbf{x}$, and there exists a non-zero vector $\mathbf{x}$ such that $\mathbf{x}^T \mathbf{H} \mathbf{x} = 0$. [Boesch, 1984]. This fact leads to another result:

Theorem 1.26 *Let* $\mathbf{P}$ *be a real, symmetric, positive semi-definite* $n \times n$ *matrix such* $\mathbf{P1} = \mathbf{0}$. *Then, if* $\lambda_1(\mathbf{P}) \leq \lambda_2(\mathbf{P}) \leq \cdots \leq \lambda_n(\mathbf{P})$ *are the eigenvalues of* $\mathbf{P}$, *we have* $\lambda_2(\mathbf{P}) \leq \dfrac{n}{n-1} \min_i [p_{ii}]$.

Proof Observe that $\lambda_2 = \min \ \mathbf{x}^T \mathbf{P} \mathbf{x}$ with $\mathbf{x} \perp \mathbf{1}, \|\mathbf{x}\| = 1$, by Theorem 0.4. The reasoning is as follows: let $\left[\dfrac{1}{\sqrt{n}}, \mathbf{x_2}, \mathbf{x_3}, \ldots, \mathbf{x_n} \right]$ be an orthonormal basis of eigenvectors, and consider $\mathbf{x} = \displaystyle\sum_{i=2}^{n} \alpha_i \mathbf{x}_i$. Now $\mathbf{x}^T \mathbf{P} \mathbf{x} = \displaystyle\sum_{i=2}^{n} \lambda_i \alpha_i^2 \geq \lambda_2 \|\mathbf{x}\|^2$. Also, $\mathbf{x}^T \mathbf{P} \mathbf{x} = \lambda_2$. Consider the matrix $\tilde{\mathbf{P}} = \mathbf{P} - \lambda_2 \left(\mathbf{I} - n^{-1} \mathbf{J} \right)$. If $\mathbf{y} \in \mathbb{R}^n$, write $\mathbf{y} = c_1 \mathbf{1} + c_2 \mathbf{x}$, where $\mathbf{x}$ is orthogonal to $\mathbf{1}$. Then consider $\mathbf{y}^T \mathbf{P} \mathbf{y} = c_2 \mathbf{x}^T \mathbf{P} c_2 \mathbf{x}$ which equals

$$c_2^2 \left(\mathbf{x}^T \mathbf{P} \mathbf{x} - \lambda_2 \mathbf{x}^T \left(\mathbf{I}_n - n^{-1} \mathbf{J}_n \right) \mathbf{x} \right) \geq c_2^2 \left(\lambda_2 \|\mathbf{x}\|^2 - \lambda_2 \|\mathbf{x}\|^2 \right) = 0.$$

Therefore, $\mathbf{P}$ is non-negative definite, and thus,

$$\min_i \left(m_{ii} - \lambda_2 \left(1 - \frac{1}{n} \right) \right) \geq 0. \qquad \square$$

There are two immediate corollaries to this result.

Corollary 1.27 *For any multigraph M,* $\lambda_2(M) \le \dfrac{n}{n-1}\delta.$

Proof Clearly, $\lambda_2(M) \le \dfrac{n}{n-1}\min[h_{ii}] = \dfrac{n}{n-1}\delta.$ □

Corollary 1.28 *For any multigraph M,* $\lambda_n(M) \ge \dfrac{n}{n-1}\Delta.$

Proof Since $\Delta(M) = (n-1)g - \delta(\bar{M}),$

$$\delta\left(\bar{M}^{(g)}\right) = (n-1)g - \Delta(M).$$

But

$$\lambda_2\left(\bar{M}^{(g)}\right) \le \frac{n}{n-1}\delta\left(\bar{M}^{(g)}\right) = \frac{n}{n-1}\left((n-1)g - \Delta(M)\right) = ng - \frac{n}{n-1}\Delta(M).$$

Hence,

$$ng - \lambda_n(M) = \lambda_2\left(\bar{M}^{(g)}\right) \le ng - \frac{n}{n-1}\Delta(M), \text{ or } \lambda_n(M) \ge \frac{n}{n-1}\Delta(M).$$

□

1.8 Two Maximum Tree Results

It can be said that K_n has the greatest number of spanning trees of any graph in the same class. As a matter of fact, K_n is the *only* graph in its class, so the result is obvious. However, it is also true that K_n has the greatest number of spanning trees among all multigraphs in its class, and that it is unique in this regard. This was originally established by [Kelmans, 1974]. We include a proof for completeness. First, because the eigenvalues of $\mathbf{H}(K_n)$ are given by $0, n, ..., n$, Cayley's Theorem

(Proposition 1.6(a)) follows: $t(K_n) = \dfrac{1}{n}\prod_{i=2}^{n}\lambda_i = \dfrac{1}{n}n^{n-1} = n^{n-2}$. Since

the diagonal entries of the Laplacian matrix are the degrees of the nodes,

$trace(\mathbf{H}) = 2e = \sum_{i=1}^{n}\lambda_i$, which is the same for all multigraphs in a given

class. In this case, the sum is $n(n-1)$. By a basic result of optimization

theory, such products are maximized when all factors are equal. Thus, no

other multigraph with $e = \dbinom{n}{2}$ edges can have more spanning trees.

Now, prior to demonstrating uniqueness, it is first observed that two non-isomorphic graphs in the same class can have the same eigenvalues, so uniqueness of the eigenvalues is not a triviality. Two non-isomorphic graphs with the same eigenvalues are called *cospectral*. Consider the two graphs in Figure 1.14. While both are in the class $\Omega(6,9)$, and regular of degree 3, they are not isomorphic. However, calculation of the eigenvalues of their respective Laplacian matrices yields the identical sequence (0, 1.438, 3, 4, 4, 5.562).

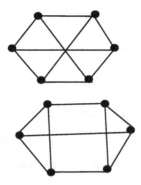

Figure 1.14: Two non-isomorphic cospectral graphs.

We prove that K_n is unique in this regard by contradiction. Suppose there exists some multigraph which has eigenvalue n with multiplicity $n-1$. Thus, the null space of $(\mathbf{H} - n\mathbf{I})$ has dimension $n-1$, or rank $(\mathbf{H} - n\mathbf{I}) = 1$. This fact, together with the symmetric nature of $\mathbf{H}$, implies that $\mathbf{H} - n\mathbf{I} = \chi\mathbf{J}$ for some constant χ. In the graph K_n, all nodes are adjacent. Any multigraph in the class will have at least one pair of nonadjacent nodes, i.e., some $h_{ij} = 0$ in $\mathbf{H}$. It follows that $\chi\mathbf{J} = 0$, or $\chi = 0$, which contradicts *rank* = 1. Hence, we have established the next theorem:

Theorem 1.29 *The complete graph K_n is the unique multigraph in its class having the greatest number of spanning trees.*

The following theorem extends this result. After obtaining this result we learned that apparently it was also proven by Kelmans [Kelmans, 1967]. However, since the original paper has never been translated into English, we shall provide a proof here, which first appeared in [Heinig, 2013]:

Theorem 1.30 *The multigraph gK_n has greatest number of spanning trees over all multigraphs in its class, and it is unique in this regard.*

Proof Indeed,

$$\hat{\mathbf{B}}(gK_n) = \mathbf{H}(gK_n) + g\mathbf{J} =$$

$$\begin{bmatrix} g(n-1) & -g & \cdots & -g \\ -g & g(n-1) & \cdots & -g \\ \vdots & \vdots & \ddots & \vdots \\ -g & \cdots & -g & g(n-1) \end{bmatrix} + \begin{bmatrix} g & g & \cdots & g \\ g & g & \cdots & g \\ \vdots & \vdots & \ddots & \vdots \\ g & g & \cdots & g \end{bmatrix} =$$

$$\begin{bmatrix} gn & 0 & \cdots & 0 \\ 0 & gn & 0 & \vdots \\ \vdots & \vdots & \ddots & \vdots \\ 0 & \cdots & 0 & gn \end{bmatrix}.$$

The eigenvalues have product $(gn)^n$. To prove uniqueness, suppose that M is a different multigraph, and that $\hat{\mathbf{B}}(M)$ has n eigenvalues equal to gn. Then there is an orthonormal matrix such that

$$\mathbf{S}^{-1}\hat{\mathbf{B}}(M)\mathbf{S} = gn\mathbf{I} = gn\mathbf{S}\mathbf{S}^{-1}. \text{ Then } \mathbf{S}^{-1}\hat{\mathbf{B}}(M)\mathbf{S} = gn\mathbf{I} \Rightarrow$$

$$\mathbf{S}\mathbf{S}^{-1}\hat{\mathbf{B}}(M)\mathbf{S}\mathbf{S}^{-1} = \mathbf{S}gn\mathbf{I}\mathbf{S}^{-1} \Rightarrow \hat{\mathbf{B}}(M) = \mathbf{S}gn\mathbf{I}\mathbf{S}^{-1} = gn\mathbf{I}\mathbf{S}\mathbf{S}^{-1} = gn\mathbf{I}.$$

Therefore, $\hat{\mathbf{B}}(M) = ng\mathbf{S}\mathbf{S}^{-1} = ng\mathbf{I}$, and $M = gK_n$.   □

Remark 1.8 Of course, this subsumes the case $1K_n = K_n$ and provides a much simpler proof than that given previously.

Chapter 2

Multigraphs with the Maximum Number of Spanning Trees: An Analytic Approach

In this chapter, we consider the optimization problem of maximizing the number of spanning trees among all multigraphs in $\Omega(n,e)$, i.e., among all multigraphs having n nodes and e edges, for certain values of e. Since a spanning tree is a minimally connected subgraph, multigraphs having more of these are, in some sense, more immune to disconnection by edge failure.

A result of Cheng [Cheng, 1981] in optimization theory states that a product of positive elements having fixed sum A and fixed sum of squares C is maximized when the elements are all the same or "almost" the same. He made use of this result to find the multigraph that maximizes the number of spanning trees for two different values of e, and the following is an exposition and expansion of his treatment.

2.1 The Maximum Spanning Tree Problem

With $t(M)$ denoting the number of spanning trees of the multigraph M, we define the *maximum spanning tree problem*, $\mathscr{P}_t$, by

Problem $\mathcal{P}_t$:

Maximize $t(M)$ subject to

$$M \in \Omega(n,e).$$

Since by Theorem 1.5, the number of spanning trees of a multigraph M is related to the eigenvalues $0 = \lambda_1 \leq \lambda_2 \leq ... \leq \lambda_n$ of its Laplacian matrix $\mathbf{H}(M)$ as follows: $t(M) = \dfrac{1}{n} \prod\limits_{i=2}^{n} \lambda_i(M),$ we observe that the following problem is a relaxation of $\mathcal{P}_t$,

Problem $\mathcal{P}_h$:

Maximize $\dfrac{1}{n} \prod\limits_{i=2}^{n} \lambda_i$ subject to

1. $\{0, \lambda_1, \lambda_2, ..., \lambda_n\}$ is the spectrum of an $n \times n$ symmetric matrix $\mathbf{H} = \left[h_{ij} \right]$ of integers, with each $h_{ij} \leq 0$ for $i \neq j$.

2. $\sum\limits_{j=1}^{n} h_{ij} = 0$ for each i.

3. $\sum\limits_{i=1}^{n} h_{ii} = 2e$.

Note that this relaxation contains instances which are not feasible for $\mathcal{P}_t$ because, if $\mathbf{H} = \left[h_{ij} \right]$ is the Laplacian of a multigraph M then the main diagonal terms must also meet the condition: if $h_{kk} = \max\limits_{1 \leq i \leq n} h_{ii}$ then

$h_{kk} \leq \sum_{i \neq k} h_{ii}$ [Hakimi, 1962]. The reader will also observe that both $\mathscr{P}_t$ and $\mathscr{P}_h$ have finitely many feasible solutions. Since $\mathbf{H}$ is a real, symmetric, positive semi-definite matrix all λ_i's ≥ 0, by Proposition 0.7 and trace $(\mathbf{H}) = \sum_{i=2}^{n} \lambda_i$, by Theorem 0.11. Thus the problem

Problem $\mathscr{P}_s$:

Maximize $\dfrac{1}{n} \displaystyle\prod_{i=2}^{n} \lambda_i$ subject to

1. $\lambda_i \geq 0$ for all $i = 2, 3, \ldots, n$

2. $\displaystyle\sum_{i=2}^{n} \lambda_i = 2e$,

is a relaxation of $\mathscr{P}_h$ with infinitely many instances.

A sometimes fortuitous approach to solving a constrained optimization problem is to obtain a relaxation that has a readily obtainable solution which is feasible for the original problem. Even though $\mathscr{P}_s$ is a relaxation of $\mathscr{P}_t$ we are not able to use this simple approach in this case. Rather, we add an additional constraint to both $\mathscr{P}_h$ and $\mathscr{P}_s$ of the form $\displaystyle\sum_{i=2}^{n} \lambda_i^2 = C$. Realize that this condition can be written as

$$\sum_{i=1}^{n} h_{ii}^2 + \sum_{i \neq j} h_{ij}^2 = C$$

in the context of $\mathcal{P}_h$. We refer to the newly obtained problems as $\mathcal{P}_{h,C}$
and $\mathcal{P}_{s,C}$ and impose conditions on C and e that guarantee the existence
of a solution obtainable by the Lagrange multiplier technique. Broadly
speaking, we determine that the value of the optimal solution of $\mathcal{P}_{s,C}$
increases as C decreases. Thus by determining C_m, the minimum C over
the finite collection of C's which yield feasible solutions for $\mathcal{P}_{h,C}$, we
obtain a largest optimal for such a problem. Since each instance of $\mathcal{P}_t$ is
feasible for one of the $\mathcal{P}_{h,C}$'s, if there is an instance of $\mathcal{P}_t$ that is feasible
for $\mathcal{P}_{h,C_m}$ then a solution for $\mathcal{P}_t$ is obtained. As it happens there are two
such cases that we shall present.

To begin with we give necessary and sufficient conditions on C and e
to insure that $\mathcal{P}_{s,C}$ is well-posed, i.e. has feasible solutions and an
optimal objective value.

Lemma 2.1 *Problem $\mathcal{P}_{s,C}$ is well-posed and has a solution if and only if*

$$\frac{4e^2}{n-1} \le C \le 4e^2.$$ *Furthermore, the set of feasible solutions contains an n-*

tuple with all positive entries if and only if $\dfrac{4e^2}{n-1} \le C < 4e^2.$

Proof First suppose that $\lambda_2, \lambda_3, \ldots, \lambda_n$ satisfies the constraints. Then, with
$\lambda = (\lambda_2, \lambda_3, \ldots, \lambda_n)$ and $\mathbf{1} = (1,1,\ldots,1)$, the Cauchy-Schwarz Inequality
forces $2e = \lambda \cdot \mathbf{1} \le |\lambda||\mathbf{1}| = \sqrt{C}\sqrt{n-1}$. Thus $\dfrac{4e^2}{n-1} \le C$. Of course, since

each $\lambda_i \ge 0$ and $4e^2 = \left(\displaystyle\sum_{i=2}^{n} \lambda_i\right)^2$ it follows that $4e^2 = C + 2\displaystyle\sum_{i<j} \lambda_i \lambda_j \ge C$ and

so $\dfrac{4e^2}{n-1} \leq C \leq 4e^2$. Moreover, if one of the λ_i's is positive, $4e^2 > C$.

Conversely, suppose the inequalities hold. In the event that $C = 4e^2$, the $n-1$-tuple $(2e,0,0,\ldots,0)$ satisfies the constraints. Next suppose that

$\dfrac{4e^2}{n-1} \leq C < 4e^2$. Realize that $\lambda = \dfrac{2e}{n-1}\mathbf{1}$ terminates on the plane

$\lambda \cdot \mathbf{1} = 2e$. Now $\left[(2e,0,0,\ldots,0) - \dfrac{2e}{n-1}\mathbf{1}\right] \cdot \left[\dfrac{2e}{n-1}\mathbf{1}\right] = 0$ so $x = \dfrac{2e}{n-1}\mathbf{1}$

and $y = \sqrt{C - \dfrac{4e^2}{n-1}}\,\mathbf{u}$, where $\mathbf{u}$ is the unit vector in the direction of

$(2e,0,0,\ldots,0) - \dfrac{2e}{n-1}\mathbf{1}$, are perpendicular. We claim that $\lambda = x + y$ is a

vector with positive components that satisfy the constraints of $\mathscr{P}_{s,C}$.
Indeed,

$$\lambda \cdot \mathbf{1} = x \cdot \mathbf{1} + y \cdot \mathbf{1} = x \cdot \mathbf{1} = 2e.$$

Also

$$|\lambda|^2 = \left(\dfrac{2e}{\sqrt{n-1}}\right)^2 + C - \dfrac{4e^2}{n-1} = C.$$

Next

$$\left|(2e,0,\ldots,0) - \left(\dfrac{2e}{n-1},\ldots,\dfrac{2e}{n-1}\right)\right| = 2e\left|\left(1 - \dfrac{1}{n-1}, -\dfrac{1}{n-1},\ldots,-\dfrac{1}{n-1}\right)\right| = 2e\sqrt{1 - \dfrac{1}{n-1}}$$

so that $\mathbf{u} = \dfrac{1}{\sqrt{1 - \dfrac{1}{n-1}}}\left(1 - \dfrac{1}{n-1}, -\dfrac{1}{n-1},\ldots,-\dfrac{1}{n-1}\right)$. Therefore

$$\lambda_2 = \dfrac{2e}{n-1} + \sqrt{C - \dfrac{4e^2}{n-1}}\sqrt{1 - \dfrac{1}{n-1}} > 0$$

and, for $i \geq 3$

$$\lambda_i = \frac{2e}{n-1} - \sqrt{C - \frac{4e^2}{n-1}} \left(\frac{1}{\sqrt{1 - \frac{1}{n-1}}} \frac{1}{n-1} \right) > 0.$$

Of course the continuous function $\frac{1}{n}\prod_{i=2}^{n}\lambda_i$ achieves a maximum value

on the compact set defined by the constraints.   □

Remark 2.1 If $\frac{4e^2}{n-1} \le C \le 4e^2$ holds, then Problem $\mathscr{P}_{s,C}$ has an optimal

solution, because the objective function is continuous on the nonempty
compact set defined by the constraints.

Since the optimal solution of Problem $\mathscr{P}_{s,C}$ has positive entries

whenever $\frac{4e^2}{n-1} \le C < 4e^2$, replacement of constraint (1) in $\mathscr{P}_{s,C}$ by the

requirement that all λ_i be strictly positive, yields a problem having the

solution. We designate this problem as Problem $\mathscr{P}_{s,C,p}$:

Problem $\mathscr{P}_{s,C,p}$

Maximize $\frac{1}{n}\prod_{i=2}^{n}\lambda_i$ subject to

1. $\lambda_i > 0, \ i = 2,3,...,n$

2. $\sum_{i=2}^{n}\lambda_i = 2e$

3. $\sum_{i=2}^{n}(\lambda_i)^2 = C$.

Now, (1) of $\mathscr{P}_{s,C,p}$ defines an open set of $\mathbb{R}^{n-1}$. Recall that the LaGrange multiplier method applies to problems of the following type: those with a continuously differentiable objective function and continuously differentiable equality constraints, all defined on an open subset of $\mathbb{R}^{n-1}$ which determine a Jacobian of full rank. Thus, we may determine the optimal solutions of $\mathscr{P}_{s,C,p}$ with the aid of the LaGrange multiplier technique if we also assume $C < 4e^2$ to ensure a Jacobian of full rank.

Proposition 2.2 *In the event that* $\dfrac{4e^2}{n-1} < C < 4e^2$, *and problems* $\mathscr{P}_{s,C}$ *and* $\mathscr{P}_{s,C,p}$ *have solutions at a location* $\lambda^* = (\lambda_2^*, \lambda_3^*, ..., \lambda_n^*)$, *then* λ^* *has exactly two distinct entries.*

Proof To apply the LaGrange multiplier technique, we must consider the Jacobian of the constraints, i.e., $\mathfrak{J} = \begin{bmatrix} 1 & 1 & ... & 1 \\ 2\lambda_2 & 2\lambda_3 & ... & 2\lambda_n \end{bmatrix}$. The entries are continuous on $\mathbb{R}^{n-1}$ and therefore also in a neighborhood of λ^*. Furthermore, $rank(\mathfrak{J}) < 2$ implies that all λ_i are equal, which is impossible, because $C > \dfrac{4e^2}{n-1}$. Therefore, $rank(\mathfrak{J}) = 2$.

In addition, $f(\lambda) = \prod_{i=2}^{n} \lambda_i$ has continuous partials in a neighborhood of λ^* so that, if we consider the function

$$L(\lambda, \alpha_1, \alpha_2) = \prod_{i=2}^{n} \lambda_i + \alpha_1 \left(\sum_{i=2}^{n} \lambda_i - 2e \right) + \alpha_2 \left(\sum_{i=2}^{n} \lambda_i^2 - C \right)$$

then there exists α_1^*, α_2^* such that L has a critical point at $(\lambda^*, \alpha_1^*, \alpha_2^*)$.

In particular, $\dfrac{\partial L}{\partial \lambda_i} = \prod_{j \neq i} \lambda_j + \alpha_1 + 2\alpha_2 \lambda_i = 0$ for $i = 2,...,n$ must be

satisfied by $(\lambda^*, \alpha_1^*, \alpha_2^*)$. Suppose there exists $\lambda_{i_1}^* < \lambda_{i_2}^* < \lambda_{i_3}^*$ (i.e., three

distinct entries). Then

1. $\dfrac{\partial L}{\partial \lambda_{i_1}^*} = \prod_{j \neq i_1} \lambda_j^* + \alpha_1 + 2\alpha_2 \lambda_{i_1}^* = 0$,

2. $\dfrac{\partial L}{\partial \lambda_{i_2}^*} = \prod_{j \neq i_2} \lambda_j^* + \alpha_1 + 2\alpha_2 \lambda_{i_2}^* = 0$,

3. $\dfrac{\partial L}{\partial \lambda_{i_3}^*} = \prod_{j \neq i_3} \lambda_j^* + \alpha_1 + 2\alpha_2 \lambda_{i_3}^* = 0$.

Subtracting the first two equations leads to $\alpha_2 = \dfrac{1}{2} \prod_{j \neq i_1, i_2} \lambda_j^*$, while

subtracting the last two equations yields $\alpha_2 = \dfrac{1}{2} \prod_{j \neq i_2, i_3} \lambda_j^*$, and putting

these together implies that $\lambda_{i_1}^* = \lambda_{i_3}^*$, which contradicts $\lambda_{i_1}^* < \lambda_{i_3}^*$. Therefore,

there are only two distinct entries. $\square$

Proposition 2.3

*(i) The collection of m-tuples λ in which there are only two distinct
entries and which satisfy*

    1. $\lambda \cdot \tilde{1} = 2e$

2. $\|\lambda\|^2 = C$, $\lambda_i > 0$ *(where* $\left(where \dfrac{4e^2}{n-1} < C < 4e^2 \right)$)

is finite in number. The possible multiplicities for the larger entry L are
$1,2,\ldots,n-2$. *For each such multiplicity k,*

$$L(k,2e,C) = \frac{2e + \sqrt{\left(\dfrac{(n-1)(n-1-k)}{k}\right)\left(C - \dfrac{4e^2}{n-1}\right)}}{n-1} \quad and$$

$$S(k,2e,C) = \frac{2e - \sqrt{\left(\dfrac{(n-i)k}{n-1-k}\right)\left(C - \dfrac{4e^2}{n-1}\right)}}{n-1}$$

for the larger and smaller entries $L(k,2e,C)$ *and* $S(k,2e,C)$, *respectively, or simply, just L and S.*

(ii) The objective function $\dfrac{1}{n-1}\displaystyle\prod_{i=2}^{n}\lambda_i$ *achieves its maximum only at* $k=1$, *i.e., the solution of Problem* $P_{s,C}$ *is uniquely attained at* $\lambda^* = (L(1,2e,C),S(1,2e,C),\ldots,S(1,2e,C))$. *This location is unique up to permutation of the entries.*

Proof (i) Consider the system: $\begin{aligned} 2e &= kL + (n-1-k)S \\ C &= kL^2 + (n-1-k)S^2. \end{aligned}$

Substituting $S = \dfrac{2e - kL}{n-1-k}$ from the first equation into the second leads

to $(n-1)kL^2 - 4ekL + [4e^2 + C(k-n+1)] = 0$ and application of the quadratic formula and simplification leads to the formula for L and S.

(ii) Consider the system as before:
$$2e = kL + (n-1-k)S$$
$$C = kL^2 + (n-1-k)S^2.$$

Taking partials and solving yields
$$\frac{\partial S}{\partial k} = \frac{S-L}{2(n-1-k)}$$
$$\frac{\partial L}{\partial k} = \frac{S-L}{2k}.$$

Because there are two distinct entries, k cannot equal zero, and so 1 is the smallest possible value for k. Given $P = S^{n-1-k}L^k$, logarithmic differentiation leads to the formula $\dfrac{dP}{dk} = P\left[\ln\left(\dfrac{L}{S}\right) - \dfrac{L}{2S} + \dfrac{S}{2L}\right]$.

The substitution $x = \dfrac{L}{S}$ into $f(x) = \ln(x) - \dfrac{x}{2} + \dfrac{1}{2x}$ yields the quantity in the bracketed part of $\dfrac{\partial P}{\partial k}$. Since $f(1) = 0$ and $f' < 1$ for $x > 1$, the value within the bracket is always negative, and therefore, P is a strictly decreasing function of k. Hence, the maximum value for P is attained only when $k = 1$.   □

Proposition 2.4 *The optimal value of the objective function* $\dfrac{1}{n}\displaystyle\prod_{i=2}^{n}\lambda_i$ *is strictly decreasing as a function of C (with 2e necessarily held fixed).*

Proof It suffices to show that the first derivative of $P = S^{n-2}L$ is always negative. Indeed, $\dfrac{\partial P}{\partial C} = \dfrac{S^{n-3}}{2n-2}\left[\dfrac{L}{S-2e} + \dfrac{S(n-2)}{L-2e}\right] =$

$$\frac{S^{n-3}}{2n-2}\left[\frac{L(L-2e)+S(m-1)(S-2e)}{(S-2e)(L-2e)}\right]<0. \qquad \square$$

Before applying this to the specific problem, we require a lemma:

Lemma 2.5 *(i) The function* $L(k,2e,C)$ *is increasing as a function of C with k and 2e fixed, and decreasing as a function of k with 2e and C fixed.*

(ii) The function $P=L^{k}S^{m-k}$ *is decreasing as a function of k with 2e and C held fixed, and decreasing as a function of C with k and 2e fixed.*

Proof (i) We know

$$L=\frac{2e+\sqrt{\left(\dfrac{(n-1)(n-1-k)}{k}\right)\left(C-\dfrac{4e^{2}}{n-1}\right)}}{n-1}=$$

$$\frac{2e+\sqrt{\left(\dfrac{C(n-1)(n-1-k)}{k}\right)\left(\dfrac{4e^{2}(n-1-k)}{k}\right)}}{n-1}.$$

So, $\left(L-\dfrac{2e}{n-1}\right)^{2}=\dfrac{C(n-1-k)}{(n-1)k}-\dfrac{4e^{2}(n-1-k)}{k(n-1)^{2}}$, and the partial

derivative with respect to C is $\dfrac{\partial L}{\partial C}=\dfrac{(n-1)(n-1-k)}{2k(L(n-1)-2e)}>0.$ Similarly,

$$\frac{\partial L}{\partial k}=\frac{4e^{2}-C}{\left(k(n-1)^{2}\right)\left(2\left(L-\dfrac{2e}{n-1}\right)\right)}<0.$$

(ii) The proof of the first part is the same as that for Proposition (3.3(ii)). For the second part, consider $\dfrac{\partial L}{\partial C} = \dfrac{(n-1)(n-1-k)}{2k(L(n-1)-2e)}$

and $\dfrac{\partial S}{\partial C} = \dfrac{((n-1)k)}{2(n-1-k)(S(n-1)-2e)}$. The partial derivative

$\dfrac{\partial P}{\partial C} = L^{k-1}S^{n-k-2}\left[kS\dfrac{dL}{dC} + L\left((n-1-k)\dfrac{dS}{dC}\right)\right]$ can be shown to be

negative, as the quantity in the brackets reduces to

$\dfrac{S(n-1)(n-1-k)(S(n-1)-2e) + Lk(n-1)(L(n-1)-2e)}{2(L(n-1)-2e)(S(n-1)-2e)}$, which has

positive numerator and negative denominator. □

The following lemma is also needed.

Lemma 2.6 Consider the following problem:

Minimize $\displaystyle\sum_{i=1}^{k} x_i^2$ subject to

    1. $\displaystyle\sum_{i=1}^{k} x_i = I$

    2. x_i is a nonnegative integer for each $i = 1, 2, ..., k$.

Then there exists integers $s \geq 0$ and $1 \leq l \leq k$ such that

$$x_i = \begin{cases} s+1, & i=1,...,l \\ s, & i=l+1,...,k \end{cases}$$ is a location of the solution to this problem.

Furthermore, this location is unique up to permutation of the x_i's.

Proof Suppose $x_1, x_2, \ldots, x_k$ is feasible, and there exists a and b such

that $x_a - x_b \geq 2$. Then $\displaystyle\sum_{i=1}^{k} x_i^2 > \sum_{i \neq a,b} x_i^2 + (x_a - 1)^2 + (x_b + 1)^2$. Thus, since

there are only finitely many feasible solutions so that the minimum

$\displaystyle\sum_{i=1}^{k} x_i^2$ exists, any optimal solution can have at most two distinct entries.

Of course, the values of the larger (l) and smaller (s) subject to $I = sk + l$

are uniquely determined by the division algorithm, where $0 \leq l \leq k-1$.

If s divides I, then $l = 0$, which implies that all entries are equal. If s

does not divide I, then $sk + l = I$, so $s(k-1) + l(s+1) = I$. $\qquad\qquad\square$

2.2 Two Maximum Spanning Tree Results

Recall that trace of a matrix is the sum of its eigenvalues, and the
eigenvalues of the square of a matrix are the squares of the individual
eigenvalues. We use these observations and Lemma 2.6 to prove the next
two spanning tree results:

Theorem 2.7 *The graph* K_n^{+x}, *i.e.* K_n *with one edge doubled, has the
greatest number of spanning trees among all multigraphs in its class.*

Proof First, we note that there are no graphs in this particular class. In
order to prove the theorem, we require the eigenvalues for the Laplacian
matrix of K_n^{+x}.

Claim: The   Laplacian   matrix   of K_n^{+x} has   eigenvalues $\lambda = (0, n, n, ..., n, n + 2)$.

Proof of Claim: Without loss of generality assume the edge x is added between v_1 and v_2. The characteristic polynomial of

$$\mathbf{B}\left(K_n^{+x}\right) = \begin{bmatrix} n+1 & -1 & 0 & 0 & \cdots \\ -1 & n+1 & 0 & 0 & \cdots \\ 0 & 0 & n & 0 & \cdots \\ \cdot & \cdot & \cdot & \cdot & \cdot \\ \cdot & \cdot & \cdot & \cdot & \cdot \\ \cdot & \cdot & \cdot & \cdot & \cdot \\ 0 & 0 & 0 & \cdots & n \end{bmatrix}$$

is easily seen to be $(\lambda - n)^{n-1}(\lambda - (n+2))$. Therefore, by Corollary 1.14, the Laplacian matrix has eigenvalues $\lambda = (0, n, ..., n, n+2)$, which proves the claim.

Thus, $2e = n(n-2) + n + 2 = n^2 - 2n + n + 2 = n^2 - n + 2$, and

$$C = trace[(\mathbf{H}(K_n^{+x}))^2] = (n-2)n^2 + (n+2)^2 =$$
$$n^3 - 2n^2 + n^2 + 4n + 4 = n^3 - n^2 + 4n + 4,$$

it follows that $\dfrac{4e^2}{n-1} = \dfrac{(n^2 - n + 2)^2}{n-1} = n^3 - n^2 + 4n + \dfrac{4}{n-1} < C$.   Hence,

by Proposition 2.3(ii), it follows that $\lambda = (n+2, n, ..., n)$ is the location of the solution to Problem $\mathscr{P}_{s,C,p}$.

Next,   we   show   that   amongst   all   the   problems $\mathscr{P}_{s,C,p}$ where

$A = 2e = \dbinom{n}{2} + 1$, $C = \left(trace\left(\mathbf{H}^2(M)\right)\right)$,   and $M \in \Omega\left(n, \dbinom{n}{2} + 1\right)$,   (i.e.,

forcing a multigraph situation) the one with the one with the largest optimal solution has $C = trace(\mathbf{H}^2(K_n^{+x}))$. We do this by applying Lemma 2.5 after proving that $\min\left(trace\left(\mathbf{H}^2(M)\right)\right)$ occurs only at $M = K_n^{+x}$. Indeed, $\left(trace\left(\mathbf{H}^2(M)\right)\right) = \sum_{i=1}^{n} h_{ii} + \sum_{i \neq j}(h_{ij})^2$. It is easy to see that only the multigraph $M \in \Omega\left(n, \binom{n}{2}+1\right)$ which has two distinct values of h_{ii} that differ by one is $M = K_n^{+x}$.

Now $\sum_{i=1}^{n} h_{ii} = n^2 - n + 2$ for all $M \in \Omega\left(n, \binom{n}{2}+1\right)$ and

$\sum_{i=1}^{n} h_{ii} = \sum_{i \neq j}(-h_{ij})$. Since K_n^{+x} has two distinct values of h_{ij}, $i \neq j$, which differ by one, K_n^{+x} minimizes $\sum_{i \neq j}(h_{ij})^2 = \sum_{i \neq j}(-h_{ij})^2$ over all

$M \in \Omega\left(n, \binom{n}{2}+1\right)$. $\square$

Next, we consider the class of multigraphs $\Omega\left(2q, q^2\right)$. Cheng showed that $K_{q,q}$, the complete bipartite graph, has the most spanning trees over all multigraphs in this class. We include a proof for completeness:

Theorem 2.8 *The graph* $K_{q,q}$ *is the unique multigraph having the greatest number of spanning trees among all multigraphs in its class.* *[Cheng 1981]*

Proof As in Theorem 2.7, it suffices to show that $trace(\mathbf{H}(K_{p,p}))^2 <$ $trace(\mathbf{H}(M))^2$ for every multigraph M in its class. However, since $trace(\mathbf{B}(M))^2 = trace(\mathbf{H}(M))^2 + n^2$, it also suffices to prove that $K_{q,q}$ uniquely minimizes $trace(\mathbf{B}^2)$ over all $M \in \Omega\left(2q, q^2\right)$.

Consider $trace(B(G))^2 = \sum_{i=1}^{n} b_{ii}^2 + \sum_{i \neq j} b_{ij}^2$. Since $\sum_{i=1}^{n} b_{ii} = 2q^2 + 2q$, and $K_{q,q}$ has all $b_{ii} = q + 1$, it follows, as in the proof of Theorem 2.7, that $K_{q,q}$ minimizes $\sum_{i=1}^{n} b_{ii}^2$ over $\Omega\left(2q, q^2\right)$.

Next realize that $\sum_{i \neq j} b_{ij} = 2\binom{2q}{2} - 2q^2$ for all $M \in \Omega(2q, q^2)$. Furthermore, the off-diagonal entries of $B(K_{q,q})$ are 0's and 1's, so that it follows, as in the proof of Theorem 2.7, that $\sum_{i \neq j} b_{ij}$ is minimized by $K_{q,q}$. Hence, $K_{q,q}$ has a maximum number of spanning trees in $\Omega(2q, q^2)$.

To see that $K_{q,q}$ is the unique such multigraph in $\Omega(2q, q^2)$, first we claim that no multigraph M which is not a graph can have the same number of spanning trees as $K_{q,q}$. Indeed, as there are at least three distinct off-diagonal entries of $\mathbf{B}(M)$, it follows by Lemma 2.6 that

$trace(\mathbf{B}(M))^2 > trace(\mathbf{B}(K_{q,q}))^2$. Hence, it follows, as in the proof of Theorem 2.7, that $t(M) < t(K_{q,q})$.

Next, we claim that no other graph G in $\Omega(2q, q^2)$ can have the same number of spanning trees as $K_{q,q}$. If $(\mathbf{H}^2(G)) > trace(\mathbf{H}^2(K_{q,q}))$, then as above, $t(G) < t(K_{q,q})$. Hence, it remains to consider the case where $(\mathbf{H}^2(G)) = trace(\mathbf{H}^2(K_{q,q}))$, i.e., G provides an instance of the problem $\mathscr{P}_{s,C,p}$, which has a unique (up to permutation) solution at $(0, q, q, \ldots, q, 2q)$ by Proposition 2.3.

We claim that the eigenvalues of $\mathbf{H}(G)$ do not equal $0, q, q, \ldots, q, 2q$. On the contrary, suppose that the eigenvalues of G are $0, q, q, \ldots, q, 2q$. Then the eigenvalues of $\overline{G}$ are $0, 0, q, \ldots, q$ by Theorem 1.20. Hence, $\overline{G}$ has two components (see the note following Theorem 1.5), say G_1, G_2. Since the maximum eigenvalue of a graph cannot exceed the number of nodes in the graph, we have $q \le n(G_1)$ and $q \le n(G_2)$. However, $n(G_1) + n(G_2) = 2q$, so $n(G_1) = n(G_2) = q$, and the eigenvalues of G_1 and G_2 are $0, p, p, \ldots, p$. We know from the proof of Theorem 1.27 that only K_q has these eigenvalues. Thus, $\overline{G} = K_q \cup K_q$, and the contradiction $G = K_{q,q}$ follows. $\qquad\qquad\square$

Remark 2.2 Cheng mentioned but did not prove uniqueness of the eigenvalues for these results.

Chapter 3

Threshold Graphs

3.1 Characteristic Polynomials of Threshold Graphs

Recall that, if the nodes of G are perfectly reliable, edges operate independently all with the same probability p and ς_i equals the number of spanning connected subgraphs having i edges, then the All-Terminal Reliability (ATR) of a graph G can be defined by

$$R(G) = \sum_{i=n-1}^{e} \varsigma_i p^i (1-p)^{e-i}.$$ For small values of p, the reliability

polynomial will be dominated by the ς_{n-1} term, and since ς_{n-1} is the number of spanning trees, graphs with a larger number of spanning trees will have greater reliability for such p. Computing the ATR of a probabilistic graph has been demonstrated to be NP-hard [Ball, 1982]. Thus, it is advantageous to bound from below the ATR of a network represented by a graph in a particular class $\Omega(n,e)$ (i.e., all graphs on n nodes and e edges). It has been conjectured that the best such bound is provided by a threshold graph [Petingi, 1996]. If a graph minimizes the number of spanning trees, it will minimize reliability for small values of p. Let $N(u)$ denote the *deleted neighborhood* of node u, i.e. the set of all nodes adjacent to u. A graph G is a *threshold graph* if, for all pairs of nodes u and v in G,

$N(u) - \{v\} \subseteq N(v) - \{u\}$ whenever $\deg(u) \leq \deg(v)$.   An   alternate equivalent description of a threshold graph is given in our next result. We present it without proof.

Proposition 3.1 *[Hammer, 1978] A graph G on n nodes is a threshold graph if and only if the node set V of G is the disjoint union of sets U and W such that $\langle U \rangle = K_c$ and $\langle W \rangle = (n-c)K_1$, and if $u, v \in W$ such that $\deg(u) \leq \deg(v)$, then $N(u) \subseteq N(v)$. In other words, G consists of a clique and an independent set of nodes such that the neighborhoods of the nodes of the independent set are "nested in the clique."*

For convenience, the node set of threshold graphs can be thought of as having two parts: one that induces a clique, and the other independent nodes which are adjacent to subsets of the clique in such a way that their neighborhoods are nested within each other (see Figure 3.1).

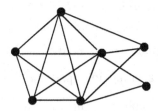

Figure 3.1: A threshold graph - K_5 is the clique, and the two independent
nodes have degrees 2 and 3, respectively.

It is somewhat easier to visualize a threshold graph if we depict it differently. We put the nodes corresponding to the clique in a box and suppress depicting the edges. Thus the only edges shown are those from the independent set of vertices to nodes in the clique.

Using this convention, Figure 3.2 depicts the same threshold graph as that given in Figure 3.1. We will use this depiction in the sequel.

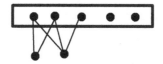

Figure 3.2: A threshold graph - K_5 is the clique (nodes in the rectangular box), and the two independent nodes have degrees 2 and 3, respectively.

Because of the way the independent nodes and the edges incident on them are depicted we refer to the independent set of nodes as *cone points*.

In the following discussion, we prove the validity of a formula for the characteristic polynomial of the Laplacian matrix of a threshold graph in terms of its degree sequence. In preparation for the proof, we require two lemmas. Recall $P_{\mathbf{H}(M)}(\lambda)$ is the characteristic polynomial of $\mathbf{H}(M)$.

Lemma 3.2 *If G has an isolated node u, then* $P_{\mathbf{H}(G)}(\lambda) = \lambda P_{\mathbf{H}(G-u)}(\lambda)$.

Proof Label the nodes of G $v_1, v_2, \ldots, v_n$ so that $u = v_1$. Since u is an isolated node, the first row of the Laplacian $\mathbf{H}(G)$ is equal to $\mathbf{0}^T$ and $\mathbf{H}_{11}(G) = \mathbf{H}(G-u)$. Thus the first row of $\lambda \mathbf{I}_n - \mathbf{H}(G)$ is $[\lambda, 0, \ldots, 0]$. Evaluating $P_{\mathbf{H}(G)}(\lambda) = \det(\lambda \mathbf{I}_n - \mathbf{H}(G))$ by expansion along the first row, we obtain $P_{\mathbf{H}(G)}(\lambda) = \lambda \det(\lambda \mathbf{I}_{n-1} - \mathbf{H}_{11}(G)) = \lambda P_{\mathbf{H}_{11}(G)}(\lambda) = P_{\mathbf{H}(G-u)}(\lambda)$. $\qquad\square$

Lemma 3.3 *If G has a full degree node u, i.e., $\deg(u) = n-1$, then*

$$(\lambda - 1)P_{H(G)}(\lambda) = \lambda(\lambda - n)P_{H(G-u)}(\lambda - 1).$$

Proof Realize u is a full degree node in G if and only if it is isolated in $\overline{G}$. Thus, $P_{\mathbf{H}(\overline{G})}(\lambda) = \lambda P_{\mathbf{H}(\overline{G}-u)}(\lambda)$. However, $\overline{G} - u = \overline{G - u}$, so $P_{\mathbf{H}(\overline{G})}(\lambda) = \lambda P_{\mathbf{H}(\overline{G-u})}(\lambda)$. Let $0, \lambda_2, ..., \lambda_{n-1}$ be the eigenvalues of $\mathbf{H}(\overline{G - u})$ so that $0, n-1-\lambda_2, ..., n-1-\lambda_{n-1}$ are the eigenvalues of $\mathbf{H}(G - u)$. Thus $P_{\mathbf{H}(G-u)}(\lambda) = \lambda \prod_{i=2}^{n-1}(\lambda - (n-1-\lambda_i))$ and so,

$P_{\mathbf{H}(G-u)}(\lambda - 1) = (\lambda - 1)\prod_{i=2}^{n-1}(\lambda - (n-\lambda_i))$. Now, the eigenvalues of $\mathbf{H}(\overline{G})$ are $0, 0, \lambda_2, ..., \lambda_{n-1}$ so that those of $\mathbf{H}(G)$ are $0, n, n-\lambda_2, ..., n-\lambda_{n-1}$ and $P_{\mathbf{H}(G)}(\lambda) = \lambda(\lambda - n)\prod_{i=2}^{n-1}(\lambda - (n - \lambda_i))$. Thus, $(\lambda - 1)P_{\mathbf{H}(G)}(\lambda) = \lambda(\lambda - n)P_{\mathbf{H}(G-u)}(\lambda - 1)$, and the proof is complete.   □

We are now ready to state the formula for the characteristic polynomial $P_{\mathbf{H}(G)}(\lambda)$ where G is a threshold graph and prove the validity of the formula. The formula has appeared in several publications [Bapat, 2010; Bleiler, 2007; Bogdanowicz, 1985], but we present what Paul Erdos would have termed "the proof from the book." It is based on an outline due to our friend and colleague, the late Frank T. Boesch.

Theorem 3.4 *If $d_1 \leq d_2 \leq \cdots \leq d_n$ denotes the degree sequence of a threshold graph on n nodes, and k is the number of cone nodes,* *i.e.,* $d_{k+1} = n - k - 1, k \geq 1, d_1 \geq 1$, *then the characteristic polynomial of* **H** *is given by*

$$P_{\mathbf{H}(G)}(\lambda) = \lambda(\lambda - n)^{d_1} \prod_{i=1}^{k}(\lambda - d_i)\prod_{j=1}^{k}(\lambda - n + j)^{d_{j+1}-d_j}.$$

Proof The proof proceeds by induction on $n \geq 3$. There is only one threshold graph when $n = 3$ that satisfies the hypotheses, namely, P_3, where $k = 1, d_1 = 1$. In this case, the formula yields

$$P_{\mathbf{H}(G)}(\lambda) = \lambda(\lambda - 3)^1(\lambda - 1)^1(\lambda - 2)^{1-1} = \lambda(\lambda - 3)(\lambda - 1),$$

which is equal to $\det(\lambda \mathbf{I}_3 - \mathbf{H}(P_3))$.

Next, we make the induction hypothesis that the formula is valid for threshold graphs having at most $n - 1$ but at least 3 nodes that satisfy the hypotheses. Consider a threshold graph on n nodes with degree sequence satisfying the hypotheses. Since $d_1 \geq 1$, there exists a full degree node u adjacent to the cone v_1. Removal of u produces one of the following three cases:

(i) if $d_1 > 1$, we obtain a threshold graph on $n - 1$ nodes with k cone nodes having degrees $1 \leq d_1 - 1 \leq ... \leq d_k - 1$;

(ii) if $d_i = 1$, for $1 \leq i \leq \ell < k$, and $d_{\ell+1} \geq 2$, we obtain a threshold graph on $n - \ell - 1$ nodes with $k - \ell$ cone nodes degrees $1 \leq d_{\ell+1} - 1 \leq ... \leq d_k - 1$ and ℓ isolated nodes;

(iii) if $d_1 = d_2 = ... = d_k = 1$, we obtain a complete graph K_{n-k-1} and k isolated nodes.

For case (i), $d_{k+1} - 1 = (n-1) - k - 1$. Applying the induction hypothesis, we obtain

$$P_{\mathbf{H}(G-u)}(\lambda) =$$

$$\lambda(\lambda-(n-1))^{d_1-1}\prod_{i=1}^{k}(\lambda-(d_i-1))\prod_{j=1}^{k}(\lambda-(n-1)+j)^{d_{j+1}-d_j}$$

so that

$$P_{\mathbf{H}(G-u)}(\lambda-1) = (\lambda-1)(\lambda-n)^{d_1-1}\prod_{i=1}^{k}(\lambda-d_i)\prod_{j=1}^{k}(\lambda-n+j)^{d_{j+1}-d_j}.$$

Thus, by Lemma 3.3,

$$(\lambda-1)P_{\mathbf{H}(G)}(\lambda) = \lambda(\lambda-n)P_{\mathbf{H}(G-u)}(\lambda-1) =$$

$$\lambda(\lambda-1)(\lambda-n)^{d_1}\prod_{i=1}^{k}(\lambda-d_i)\prod_{j=1}^{k}(\lambda-n+j)^{d_{j+1}-d_j},$$

and the conclusion follows after cancellation of $\lambda-1$.

For case (ii), observe that

$$(n-\ell-1)-(k-\ell)-1 = n-k-2 = d_{k+1}-1,$$

so, by Lemma 3.2 and the induction hypothesis,

$$P_{\mathbf{H}(G-u)}(\lambda) = \lambda^{\ell+1}P_{\mathbf{H}(G-u-v_1-v_2-\ldots-v_\ell)}(\lambda) =$$

$$\lambda^{\ell+1}(\lambda-(n-\ell-1))^{d_{\ell+1}-1}\prod_{i=\ell+1}^{n}(\lambda-(d_i-1))\prod_{j=\ell+1}^{k}\left(\lambda-(n-\ell-1))+j-\ell\right)^{d_{j+1}-d_j}$$

so that $P_{\mathbf{H}(G-u)}(\lambda-1) =$

$$(\lambda-1)^{\ell+1}(\lambda-(n-\ell))^{d_{\ell+1}-1}\prod_{i=\ell+1}^{n}(\lambda-d_i)\prod_{j=\ell+1}^{k}(\lambda-(n-\ell)+j-\ell)^{d_{j+1}-d_j}.$$

Now, applying Lemma 3.3, and cancelling $\lambda-1$, we obtain

$$P_{\mathbf{H}(G)}(\lambda) =$$

$$\lambda(\lambda-n)^1(\lambda-1)^{\ell}(\lambda-(n-\ell))^{d_{\ell+1}-1}\prod_{i=\ell+1}^{k}(\lambda-d_i)\prod_{j=\ell+1}^{k}(\lambda-(n-\ell)+j-\ell)^{d_{j+1}-d_j}.$$

Observe that $(\lambda-1)^{\ell}\prod_{i=\ell+1}^{k}(\lambda-d_i)=\prod_{i=1}^{k}(\lambda-d_i)$. Also, $d_{j+1}-d_j=0$ for

$j=1,...,\ell-1$ and $d_{\ell+1}-d_{\ell}=d_{\ell+1}-1$, so

$$(\lambda-(n-\ell))^{d_{\ell+1}-1}\prod_{j=\ell+1}^{k}(\lambda-(n-\ell)+j-\ell)^{d_{j+1}-d_j}=\prod_{j=1}^{k}(\lambda-n+j)^{d_{j+1}-d_j}.$$

Hence, $P_{\mathbf{H}(G)}(\lambda)=\lambda(\lambda-n)^{d_1}\prod_{i=1}^{k}(\lambda-d_i)\prod_{j=1}^{k}(\lambda-n+j)^{d_{j+1}-d_j}$, the required

conclusion.

Finally, in case (iii),

$$P_{\mathbf{H}(G-u)}(\lambda)=\lambda^k P_{\mathbf{H}(K_{n-(k+1)})}(\lambda)=\lambda^k\lambda\big(\lambda-(n-k-1)\big)^{n-k-2},\text{ so}$$

$$P_{\mathbf{H}(G-u)}(\lambda)=\lambda^k P_{\mathbf{H}(K_{n-(k+1)})}(\lambda)=\lambda^k\lambda\big(\lambda-(n-k-1)\big)^{n-k-2}.$$

Again, applying Lemma 3.3, and cancelling $\lambda-1$, we obtain

$$P_{\mathbf{H}(G)}(\lambda)=\lambda(\lambda-n)^1(\lambda-1)^k(\lambda-(n-k))^{n-k-2}.$$

Now, $(\lambda-1)^k=\prod_{i=1}^{k}(\lambda-d_i)$ and

$$\prod_{j=1}^{k}(\lambda-n+j)^{d_{j+1}-d_j}=(\lambda-n+k)^{d_{k+1}-1}=(\lambda-n+k)^{n-k-1-1},$$

thus the result follows. □

Remark 3.1 The material concerning the above formula has been submitted for publication.

Example 3.1 Consider the threshold graph depicted in Figure 3.2. For this graph, $n = 7$, $k = 2$, and the degree sequence is $\mathbf{d} = 2,3,4,4,5,6,6$. Applying Theorem 3.4, we obtain

$$P_{\mathbf{H}(G)}(\lambda) = \lambda(\lambda-7)^2(\lambda-2)(\lambda-3)(\lambda-7+1)^{3-2}(\lambda-7+2)^{4-3} =$$

$$\lambda(\lambda-7)^2(\lambda-2)(\lambda-3)(\lambda-6)(\lambda-5),$$

so the eigenvalues are 0, 2, 3, 5, 6, 7, 7, and the number of spanning trees is $\dfrac{2\cdot 3\cdot 5\cdot 6\cdot 7\cdot 7}{7} = 1260$ by Theorem 1.7.

Another way to generate the eigenvalues for the Laplacian matrix of a threshold graph can be found in [Bapat, 2010] and is based on the work of Russell Merris [Merris, 1994]. Given a degree sequence of a threshold graph, $\mathbf{d} = d_1, d_2, \cdots, d_{n-1}, d_n$, with $d_1 \geq d_2 \geq \cdots \geq d_{n-1} \geq d_n$, we form a new sequence $\mathbf{d}^* = d_1^*, d_2^*, \cdots, d_{n-1}^*, d_n^*$, where d_i^* is the number of terms in the degree sequence $\mathbf{d}$ that are greater than or equal to i. It is the case that $\mathbf{d}^*$ is in fact the sequence of the eigenvalues $\lambda_1 \geq \lambda_2 \geq \cdots \geq \lambda_{n-1} \geq \lambda_n = 0$.

Example 3.2 For the threshold graph in Figure 3.2, we reorder the degree sequence $\mathbf{d} = 6,6,5,4,4,3,2$, i.e., $d_1 = 6, d_2 = 6, d_3 = 5, d_4 = 4$, $d_5 = 4, d_6 = 3$, and $d_7 = 2$. Thus $\lambda_1 = d_1^* = 7$, $\lambda_2 = d_2^* = 7$, $\lambda_3 = d_3^* = 6$, $\lambda_4 = d_4^* = 5$, $\lambda_5 = d_5^* = 3$, $\lambda_6 = d_6^* = 2$, and $\lambda_7 = d_7^* = 0$.

In their 1996 paper, *Laplacian spectra and spanning trees of threshold graphs* [Hammer, 1996], Hammer and Kelmans present the following formula for the eigenvalues of the Laplacian matrix of a threshold graph:

(*) Let G be a connected threshold graph, and let $\mathbf{d} = (v_1^{(n_1)}, v_2^{(n_2)}, ..., v_s^{(n_s)})$

be the degree sequence of G, where $v_i \leq v_{i+1}$ *for i = 1, ..., s−1, and* n_i *is*

the multiplicity of v_i *in the degree sequence. Then*

$$P_{\mathbf{H}(G)}(\lambda) = \prod_{i=1}^{\hat{k}} (\lambda - v_i)^{n_i} \prod_{i=\hat{k}+2}^{s} (\lambda - v_i - 1)^{n_i} (\lambda - v_{\hat{k}+1} - 1)^{n_{\hat{k}+1} - 1} \ and$$

$$t(G) = \prod_{i=1}^{\hat{k}} (v_i)^{n_i} \prod_{i=\hat{k}+2}^{s-1} (v_i + 1)^{n_i} (v_{\hat{k}+1} + 1)^{n_{\hat{k}+1} - 1} (v_s + 1)^{n_s - 1},$$

where $\hat{k} = \left\lfloor \dfrac{s-1}{2} \right\rfloor$ [Hammer, 1996].

This formula does not work for every threshold graph, as demonstrated in [Bleiler, 2007]. Consider a threshold graph having a clique of order 5 and three cones of degree 4, 3, and 2, respectively, as illustrated in Figure 3.3.

Figure 3.3: A threshold graph that serves as a counterexample to the Hammer/Kelmans spanning tree formula for threshold graphs.

The degree sequence is

$$\mathbf{d} = 2, 3, 4, 4, 5, 6, 7, 7 = (2^{(1)}, 3^{(1)}, 4^{(2)}, 5^{(1)}, 6^{(1)}, 7^{(2)}),$$

so the number of distinct terms in the degree sequence is $s = 6$. According to the formula in (*), $\hat{k} = \left\lfloor \dfrac{s-1}{2} \right\rfloor = \left\lfloor \dfrac{6-1}{2} \right\rfloor = \left\lfloor \dfrac{5}{2} \right\rfloor = 2$,

and

$$P_{\mathbf{H}(G)}(\lambda) = \prod_{i=1}^{\hat{k}} (\lambda - v_i)^{n_i} \prod_{i=\hat{k}+2}^{s} (\lambda - v_i - 1)^{n_i} (\lambda - v_{\hat{k}+1} - 1)^{n_{\hat{k}+1}-1} =$$

$$\prod_{i=1}^{2} (\lambda - v_i)^{n_i} \prod_{i=4}^{6} (\lambda - v_i - 1)^{n_i} (\lambda - v_{2+1} - 1)^{n_{2+1}-1} =$$

$$\{(\lambda-v_1)^{n_1}(\lambda-v_2)^{n_2}\}\{(\lambda-v_4-1)^{n_4}(\lambda-v_5-1)^{n_5}(\lambda-v_6-1)^{n_6}(\lambda-v_3-1)^{n_3-1}\} =$$

$$\{(\lambda-2)^1(\lambda-3)^1\}\{(\lambda-5-1)^1(\lambda-6-1)^1(\lambda-7-1)^2(\lambda-4-1)^1\} =$$

$$(\lambda-2)^1(\lambda-3)^1(\lambda-6)^1(\lambda-7)^1(\lambda-8)^2(\lambda-5)^1.$$

The eigenvalues for the graph's Laplacian matrix are, by this formula, 2, 3, 5, 6, 7, 8, 8 (plus the zero eigenvalue, which does not factor into the spanning tree computations for connected graphs). Application of Theorem 1.5 indicates that this graph would have 10,080 spanning trees. In fact, by Theorem 0.11, the sum of the eigenvalues and the sum of the degree sequence terms should both be twice the number of edges. This is a 19 edge graph, so the sum of the eigenvalues given by (*) should be 38, but is actually 39, which is a contradiction. Indeed, direct calculation using the graph's specific Laplacian matrix reveals that the eigenvalues are actually 0, 2, 3, 4, 6, 7, 8, 8 and the number of spanning trees is 8064. The formula of Hammer and Kelmans can be restated as follows:

Theorem 3.5 *[Bleiler, 2007] Let G be a connected threshold graph, and let* $\mathbf{d} = (v_1^{(n_1)}, v_2^{(n_2)}, ..., v_s^{(n_s)})$ *be the degree sequence of G, where* $v_i \le v_{i+1}$ *for i = 1, ..., s – 1, and* n_i *is the multiplicity of* v_i *in the degree sequence, with* $\hat{k} = \left\lfloor \dfrac{s-1}{2} \right\rfloor.$

(a) For s odd, $P_{\mathbf{H}(G)}(\lambda) =$

$$\lambda \prod_{i=1}^{\hat{k}} (\lambda - v_i)^{n_i} \prod_{i=\hat{k}+2}^{s} (\lambda - v_i - 1)^{n_i} (\lambda - v_{\hat{k}+1} - 1)^{n_{\hat{k}+1}-1} \ and$$

$$t(G) = \prod_{i=1}^{\hat{k}} (v_i)^{n_i} \prod_{i=\hat{k}+2}^{s-1} (v_i + 1)^{n_i} (v_{\hat{k}+1} + 1)^{n_{\hat{k}+1}-1} (v_s + 1)^{n_s-1}.$$

(b) For s even, $P_{\mathbf{H}(G)}(\lambda) =$

$$\lambda \prod_{i=1}^{\hat{k}} (\lambda - v_i)^{n_i} \prod_{i=\hat{k}+2}^{s} (\lambda - v_i - 1)^{n_i} (\lambda - v_{\hat{k}+1})^{n_{\hat{k}+1}-1}$$

$$t(G) = \prod_{i=1}^{k} (v_i)^{n_i} \prod_{i=\hat{k}+2}^{s-1} (v_i + 1)^{n_i} (v_{\hat{k}+1})^{n_{\hat{k}+1}-1} (v_s + 1)^{n_s-1}.$$

Remark 3.2 The differences between the two formulas in Theorem 3.5 are based on the fact that, in threshold graphs having an odd number of distinct degree sequence terms, there is a clear distinction between the nodes in the clique and the nodes in the independent set. On the other hand, in threshold graphs having an even number of distinct degree sequence terms, the maximum degree of the cone nodes and the minimum degree of the clique nodes is the same. Furthermore, there can be more than one cone node having this common degree, but only one clique node.

In his 1985 doctoral thesis, "Spanning Trees in Undirected Graphs," Zbignew Rysard Bogdanowicz derived many interesting spanning tree formulas. He denotes a threshold graph by Q *(n, k)* where K_{n-k} is the clique and the k cones have degrees $q_1 \geq q_2 \geq ... \geq q_k$, and states the following theorem:

Theorem 3.6 *Let* $q_1, q_2, ..., q_k$ *(* $q_i \geq q_{i+1}$ *) denote the degrees of the cone nodes of a threshold graph on n nodes, Q(n,k). Then the number of spanning trees of a Q (n, k) graph is*

$$t(Q(n,k)) = q_k n^{q_k - 1}(n-k)^{n - q_1 - k - 1} \prod_{i=1}^{k-1} q_i(n-k+i)^{q_i - q_{i+1}}$$

[Bogdanowicz, 1985].

We will apply the various formulas to compute the number of spanning trees for the threshold graph in Figure 3.3:

Example 3.3 We will apply the theorems of this section to determine the Laplacian eigenvalues and/or the number of spanning trees of the threshold graph in Figure 3.3. This graph has parameter values

$$n = 8, \ k = 3, \ s = 6, \ \hat{k} = \left\lfloor \frac{5}{2} \right\rfloor = 2, \ \text{and degree sequence}$$

$$\mathbf{d} = 2, 3, 4, 4, 5, 6, 7, 7 = (2^{(1)}, 3^{(1)}, 4^{(2)}, 5^{(1)}, 6^{(1)}, 7^{(2)}).$$

(a) By Theorem 3.4,

$$P_{\mathbf{H}(G)}(\lambda) = \lambda(\lambda - n)^{d_1} \prod_{i=1}^{k} (\lambda - d_i) \prod_{j=1}^{k} (\lambda - n + j)^{d_{j+1} - d_j} =$$

$$\lambda(\lambda - 8)^2 \prod_{i=1}^{3} (\lambda - d_i) \prod_{j=1}^{k} (\lambda - 8 + j)^{d_{j+1} - d_j} =$$

$$\lambda(\lambda - 8)^2 (\lambda - 2)(\lambda - 3)(\lambda - 4)(\lambda - 7)(\lambda - 6)(\lambda - 5)^0 =$$

$$\lambda(\lambda - 8)^2 (\lambda - 2)(\lambda - 3)(\lambda - 4)(\lambda - 7)(\lambda - 6).$$

The number of spanning trees would be

$$\frac{8^2 \cdot 7 \cdot 6 \cdot 4 \cdot 3 \cdot 2}{8} = 8064.$$

(b) By Theorem 3.5(b),

$$\lambda\prod_{i=1}^{\hat{k}}(\lambda-v_i)^{n_i}\prod_{i=\hat{k}+2}^{s}(\lambda-v_i-1)^{n_i}(\lambda-v_{\hat{k}+1})^{n_{\hat{k}+1}-1}=$$

$$\lambda\prod_{i=1}^{2}(\lambda-v_i)^{n_i}\prod_{i=4}^{6}(\lambda-v_i-1)^{n_i}(\lambda-v_3)^{n_3-1}=$$

$$\lambda(\lambda-2)(\lambda-3)(\lambda-6)(\lambda-7)(\lambda-8)^2(\lambda-4).$$

Once again, the number of spanning trees would be 8064.

(c) Using the notation of Theorem 3.6, the cone point degrees
are $q_1=4,q_2=3,q_3=2$. Now, $t(Q(n,k))=$

$$q_k n^{q_k-1}(n-k)^{n-q_1-k-1}\prod_{i=1}^{k-1}q_i(n-k+i)^{q_i-q_{i+1}}$$

$$=q_3 8^{q_3-1}(8-3)^{8-q_1-3-1}\prod_{i=1}^{k-1}q_i(8-3+i)^{q_i-q_{i+1}}$$

$$=2\cdot8\cdot4\cdot6\cdot3\cdot7=8064.$$

3.2 Minimum Number of Spanning Trees

The "lollipop" graph, illustrated in Figure 3.4 has the same ATR and number of spanning trees as the threshold graph depicted in Figure 3.5. For our purposes here the placement of the bridges (a *bridge* is an edge whose removal disconnects the graph) as either a path joined to the remainder of the graph or single edges joined to the graph does not matter as all those edges must appear in every connected spanning subgraph, in particular every spanning tree.

This graph has the least number of spanning trees of any connected graph in its class [Bogdanowicz, 2009], and has been conjectured to form a lower bound on the ATR for all connected graphs in its class.

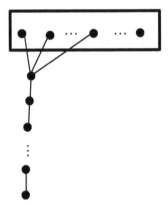

Figure 3.4: Lollipop - K_{n-k} with 1 cone of degree c and a P_{k-1} appended to the cone.

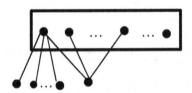

Figure 3.5: Threshold graph having K_{n-k} with 1 cone of degree c and $k-1$ cones of degree 1.

Direct application of Theorem 3.4 and Theorem 1.5 yields the number of spanning trees for a lollipop.

Proposition 3.7 *Let G be a threshold graph with clique K_{n-k}, one cone of degree c and $k-1$ cones of degree 1. Then*

$$t(G) = c(n-k+1)^{c-1}(n-k)^{n-k-1-c} .$$

Proof By Theorem 3.4 $P_{\mathbf{H}(G)}(\lambda) =$

$$\lambda(\lambda-n)(\lambda-1)^{k-1}(\lambda-c)(\lambda-n+(k-1))^{c-1}(\lambda-n+k)^{n-k-1-c}.$$

The formula follows by applying Theorem 1.5. □

We observe that the formula in Proposition 3.7 is applicable to a threshold graph in which $k=1$. Such a graph is a special subcase of the lollipop, called the balloon, depicted in Figure 3.6. When such a graph has cone degree strictly greater than 1, the balloon graph resides in a class $\Omega(n,e)$ in which connected graphs do not have any edges which are bridges. In these graphs,

$$\frac{(n-1)(n-2)}{2}+1 < e \leq \binom{n}{2} \quad (**).$$

Figure 3.6: Balloon graph having K_{n-1} with 1 cone of degree c.

As stated earlier, it has been proven [Bogdanowicz, 2009] that the lollipop minimizes the number of spanning trees among all graphs in its class. In addition, it has been conjectured to have uniformly minimum ATR among all graphs in its $\Omega(n,e)$ class. In classes of graphs whose number of edges satisfy (**), the balloon graph can be shown to minimize the number of spanning trees.

This result can be proven using a graph transformation called the swing surgery. This graph transformation is also used in the approaches to multigraphs in Chapter 4. Results regarding this surgery were established for its effect on spanning trees in the graph case by Satyanarayana, Schoppmann and Suffel [Satyanarayana,

1992] and Bogdanowitz [Bogdanowitz 1984]; its effect on a graph's reliability was investigated by Brown, Colbourn and Devitt [Brown, 1993] and Kelmans [Kelmans 1981]. Recall that if w and y are nodes, then $m(\{x,y\})$ is the multiplicity of edge $\{x,y\}$.

Theorem 3.8 (Swing Surgery) *Let **M** be a multigraph with node set V and* $w,v \in V$ *have the property that, for each* $y \in V-\{w,v\}$,

$$m(\{w,y\}) \geq m(\{y,v\}).$$

(1) *If there exists y such that* $m(\{w,y\}) > m(\{y,v\})$, *then for any such y, the multigraph* **M'**, *obtained from M by deleting one of the edges incident on w and y and inserting an additional edge incident on y and v, has at least as many spanning trees as G, i.e.* $t(M') \geq t(M)$. (Figure 3.7a)

(2) *Furthermore, if in addition to* $m(\{w,y\}) > m(\{y,v\})$, *there exists an edge incident on y and a node* $u_1 \in N(w) \cap N(v)$, *and also an edge incident on y and a node* $u_2 \in N(w) - N(v)$, *then* $t(M') > t(M)$. **[Satyanarayana, 1992]** (Figure 3.7b)

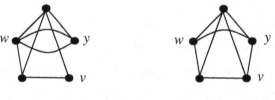

Graph G: $t(G) = 34$ Graph G': $t(G') = 45$

Figure 3.7a: Example of the swing surgery. One of the multiedges *wy* is "swung" down to become edge *yv*.

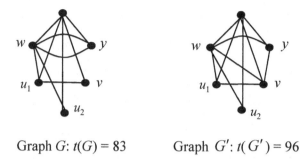

Graph G: $t(G) = 83$ Graph G': $t(G') = 96$

**Figure 3.7b: Example of the swing surgery. One of the multiedges *wy*
is "swung" down to become edge *wv*.**

We delay the proof, essentially the same as given in Schoppmann's thesis [Schoppmann, 1990], until after the presentation of three preliminary results. First we require a version of the well-known Factoring Theorem [Moskowitz, 1958] as it applies to spanning trees, as well as two lemmas:

Theorem 3.9 (Factoring Theorem) *If $M - x$ denotes the multigraph obtained by the deletion of edge x from multigraph M, and $M \mid x$ denotes the multigraph obtained by the contraction of edge x, then $t(M) = t(M - x) + t(M \mid x)$.*

The proof of this theorem is straight-forward, as the spanning trees of $M - x$ are in one-to-one correspondence with those of M that do not include x, while those of $M \mid x$ are in one-to-one correspondence with those of M that do include x.

Our next result is a lemma required to prove the theorem.

Lemma 3.10 *Let M and M' be multigraphs each having node set $V = \{w, v, y_1, y_2, \ldots, y_s\}$ with $\{y_1, y_2, \ldots, y_s\}$ an independent set of vertices. Assume further that M and M' have the same number of edges. We denote by l_i and l_i' the number of edges incident on v and y_i in M and M', respectively. Likewise, u_i and u_i' denote the number of edges incident on w and y_i in M and M', respectively. Finally, we assume that there are the same number of edges between w and v in both M and M' and let the common value of $m_M(\{w,v\}) = m_{M'}(\{w,v\})$ be denoted by h. If $|u_j - l_j| \geq |u_j' - l_j'|$ and $u_j + l_j = u_j' + l_j'$ for each $j = 1, 2, \ldots, s$, then $t(M') \geq t(M)$. Furthermore, if in addition, M' is connected and there exists an integer k such that $|u_k - l_k| > |u_k' - l_k'|$, then $t(M') > t(M)$.*

Proof (of Lemma 3.10) The proof proceeds by nested induction, first on the number of nodes in $M - w - v$, i.e., the $y_i's$, and then on the number of edges incident on y_1. (See Figure 3.8) If $s = 1$, i.e., there is exactly one y_i, then $t(M) = u_1 l_1 + h(u_1 + l_1)$ and $t(M') = u'l' + h(u' + l_1')$. It is a simple exercise to prove that $u_1' l_1' \geq u_1 l_1$ when $u_1' + l_1' = u_1 + l_1$ and $|u_1' - l_1'| \geq |u_1 - l_1|$ and that the inequality is strict provided $|u_1' - l_1'| > |u_1 - l_1|$.

Assume the lemma is true when there are $y_i's$. To prove the lemma for $s+1$ $y_i's$, we perform induction on the total number of edges incident on y_1.

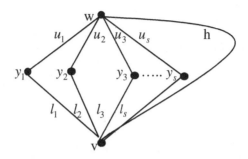

Figure 3.8: The multigraph M at the start of Lemma 3.10.
Note that nodes w and v are labeled, and the edge labels
indicate the multiplicity of each multiple edge.

Base case: Since the case $u_1 + l_1 = 0$ is trivial, i.e., $t(M) = 0 = t(M')$, assume the number of edges incident on y_1 is 1, $u_1 = 1$, and $l_1 = 0$. Without loss of generality, we can assume $u_1' = 1$ and $l_1' = 0$. Then $u_1 + l_1 = u_1' + l_1'$ and $\left| u_1' - l_1' \right| \geq \left| u_1 - l_1 \right|$. Let x_1 (x_1') be the unique edge incident on w and y_1 in M (M'). Application of the factoring theorem to x and x' yields zero spanning trees when the edge is deleted and graphs of the type described in the statement of this lemma with s y_i's when the edge is contracted. By the induction hypothesis, $t(M'|x') \geq t(M|x)$. Furthermore, if M' is connected and there exists an index k such that $|u_k - l_k| > |u_k' - l_k'|$, then necessarily, $u_1 + l_1 \geq 0$, $k \geq 2$, and after the contraction of $x_1', G'|x_1'$ remains connected, and the inequality still holds, so by the induction hypothesis $t(M'|x') > t(M|x)$.

Now assume the lemma holds for $s+1$ $y_i's$ with $u_1 + l_1 = p \geq 1$. We establish the results for $u_1 + l_1 = p + 1 \geq 2$ by considering two cases.

Case 1. Assume that $|u_1 - l_1| = |u_1' - l_1'|$. Without loss of generality, assume $u_1 = u_1', l_1 = l_1'$ and $u_1 \geq l_1$. Let x_1 be one of the edges incident on w and y_1 in M and x_1' be one of the edges incident on w and y_1 in M'. We now apply the factoring theorem to x_1 and x_1'. Their deletion makes the sum of the edges incident on y_1 equal to p in both graphs. Hence, by the induction hypothesis, $t(M - x_1) \leq t(M' - x_1')$. Contraction of x_1 and x_1' yields the multigraphs of the lemma, one with $l_1 + h$ edges joining w and v and s $y_i's$, namely, $y_2, y_3, ..., y_{s+1}$ in $M|x_1$ and the second with the analogous description for $M'|x_1'$. Hence, $t(M|x_1) \leq t(M'|x_1')$, with strict inequality if $|u_k - l_k| > |u_k' - l_k'|$ for some $k \geq 2$ by the primary induction hypothesis. An application of the factoring theorem completes the proof in this case.

Case 2. Assume that $|u_1 - l_1| > |u_1' - l_1'|$, and, without loss of generality, $u_1 > l_1, u_1' \geq l_1', u_1 > u_1', l_1 < l_1'$. Consider edges x_1 (and x_1') in M (and M') incident on w and y_1 (w' and y_1'). By the secondary induction hypothesis, $t(M - x_1) \leq t(M' - x_1')$. On the other hand, contraction of these edges yields multigraphs $M|x_1$ and $M'|x_1'$ where $l_1 < l_1'$ (see Figures 4.3 and 4.4, respectively). Repeated factoring on the edges incident on w and v in the two multigraphs yields identical contractions at each application, so it remains to compare the deleted multigraphs obtained after l_1 factorings, say $\hat{M}$ and $\hat{M}'$, where $\hat{M}$ has h edges incident on w and v while $\hat{M}'$ has $h + l_1' - l_1$ edges incident on w and v. Now the primary induction hypothesis may be invoked to conclude that the number of spanning trees of $\hat{M}'$ with $l_1' - l_1$ edges from w' and v' removed is at least as great as $t(\hat{M})$. Thus, $t(\hat{M}') \geq t(\hat{M})$. If $\hat{M}$ is connected, then it is readily seen that $\hat{M}'$ is connected as well, and since $l_1' - l_1 > 0, t(M') > t(M)$. This completes the proof. □

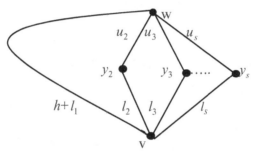

Figure 3.9: The multigraph $M \mid x_1$. Edge labels indicate the multiplicity of each multiple edge.

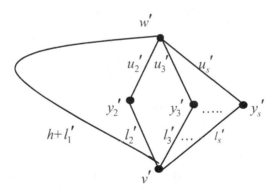

Figure 3.10: The multigraph $M' \mid x_1'$. Edge labels indicate the multiplicity of each multiple edge.

Let $x_1, x_2, ..., x_k$ be a fixed but arbitrary collection of k edges in a multigraph G. Consider edge x_1, which can be either deleted or contracted. If x_2 has become a loop, then it is deleted. (This will only happen if x_1 and x_2 are edges incident on the same pair of nodes.) If it has not become a loop, then contract or delete it. Repeat the process for x_3, and so on. We can represent any such sequence by a k-tuple **a** of 0s, 1s and dashes where

$$a_i = \begin{cases} 1 & \text{if } x_i \text{ is contracted} \\ 0 & \text{if } x_i \text{ is deleted} \\ - & \text{if } x_i \text{ has become a loop}. \end{cases}$$

A k-tuple of 0's, 1's and dashes is called *realizable* if each of its respective operations can be performed on the specified edges of a graph. We present the following lemma:

Lemma 3.11 *Let* $x_1, x_2, ..., x_k$ *be a fixed but arbitrary collection of k edges in a multigraph M.   If A consists of all such realizable k-tuples corresponding to the edges as described above, then for each* **a** $\in$ *A, let* $M_{\mathbf{a}}$ *represent the graph obtained from M by sequentially performing the operations. Then* $t(M) = \sum_{\mathbf{a} \in A} t(M_{\mathbf{a}})$.

Proof (by induction on k).   Let $k = 1$; then

$$t(M) = t(M \mid x_1) + t(M - x_1).$$

Assume the expansion is valid for $k \le m$. Now, let $k = m + 1$. By the induction hypothesis, for the first m edges, $t(M) = \sum_{\mathbf{a}' \in A'} t(M_{\mathbf{a}'})$, where

A' consists of all m-tuples corresponding to the sequential contractions and deletions of the first m edges $x_1, ..., x_m$. If for $\mathbf{a}' \in A', x_{m+1}$ is a loop in $M_{\mathbf{a}'}$, then extend $\mathbf{a}'$ by appending a dash onto $\mathbf{a}'$ and obtain $(\mathbf{a}', -) \in A$. Clearly, $t(M_{\mathbf{a}'}) = t(M_{(\mathbf{a}', -)})$. If x_{m+1} is not a loop, then $t(M_{\mathbf{a}'}) = t(M_{(\mathbf{a}', 0)}) + t(M_{(\mathbf{a}', 1)})$ by the factoring theorem. Therefore, $t(M) = \sum_{a \in A} t(M_{\mathbf{a}})$ follows. $\qquad\square$

We will now prove Theorem 3.8:

Proof of Theorem 3.8 Let $H = M - w - v = M' - w' - v'$ and have edges $x_1, ..., x_k$. We know $t(M) = \sum_{\mathbf{a} \in A} t(M_{\mathbf{a}}), t(M') = \sum_{\mathbf{a} \in A} t(M_{\mathbf{a}}')$, where A consists of the realizable k-tuples corresponding to the edges of H. Note that the hypotheses of Lemma 3.10 are satisfied so that $t(M_{\mathbf{a}}) \leq t(M_{\mathbf{a}}')$. Now suppose the conditions of Theorem 3.8 (2) hold. Arrange the edges of $M - v - w$ in an arbitrary order $x_1, x_2, ..., x_k$. Then consider the k-tuple $\mathbf{a}$ obtained by first contracting x_1 and then, for each i in turn, either contracting x_i if it is not a loop or deleting it if it is a loop. The resulting $M_{\mathbf{a}}$ and $M_{\mathbf{a}}'$ are each connected, and they satisfy the condition in Lemma 3.10 that forces $t(M_{\mathbf{a}}) < t(M_{\mathbf{a}}')$. This completes the proof of Theorem 3.8 (2). $\qquad\square$

Remark 3.3 When w and v satisfy the neighborhood conditions of Theorem 3.8(1), we say $N(w)$ *dominates* $N(v)$.

An important consequence of the Swing Surgery is that, given any connected graph G in $\Omega(n,e)$, there exists a threshold graph T also in $\Omega(n,e)$ such that $t(T) \le t(G)$. [Schoppmann, 1990] This fact, coupled with judicious application of the Swing Surgery to a threshold graph, gives the following result:

Theorem 3.12   *Consider a class of graphs $\Omega(n,e)$, such that $e \ge \dfrac{(n-1)(n-2)}{2} + 1$, that admits a unique balloon graph B and another threshold graph T having at least two cone nodes. Denote by R(G) the all terminal reliability of any graph. Then $R(B) \le R(T)$ for all values of p, i.e., the balloon is a uniform lower bound on ATR for all threshold graphs in its class. [Petingi, 1996]*

The proof of this result is performed by induction on the number of edges in the complement of the graph. A modified version of the threshold graph is transformed into a similarly modified version of the balloon graph in its class by repeated, specific application of Theorem 3.8. Consequently, since the balloon provides a uniform lower bound on ATR, it also provides a lower bound on the number of spanning trees.

3.3   Spanning Trees of Split Graphs

In section 1 of this chapter, we characterized threshold graphs as having nodes that are one of two types: nodes in a clique, and an independent set of nodes (i.e., cones) adjacent to subsets of the clique in such a way that their neighborhoods are nested within each other. If the neighborhood nesting condition is not met for at least one pair of nodes, but the graph still maintains the partition between clique and independent set, the graph is known as a *split graph*. When viewed in this scenario, a threshold graph is a special instance of the split graph, i.e., a split graph with the nesting property for every pair of nodes.

A non-increasing sequence of nonnegative integers is said to be *graphical* if there exists a graph that has the sequence as its degree sequence. Erdos and Gallai [Erdos, 1960] proved the following theorem:

Theorem 3.13 *A sequence of nonnegative integers $d_n, d_{n-1}, ..., d_1$ arranged in non-increasing order is graphical if and only if for each integer r, $1 \leq r \leq n - 1$, $\sum_{i=1}^{r} d_i \leq r(r - 1) + \sum_{i=r+1}^{n} \min\{r, d_i\}$.*

Example 3.4 Consider the sequence of integers 6, 5, 5, 5, 4, 4, 3. Now,

$$\sum_{i=1}^{1} d_i = 6 \leq 6 = 1(1-1) + (1+1+1+1+1+1);$$

$$\sum_{i=1}^{2} d_i = 11 \leq 12 = 2(2-1) + (2+2+2+2+2);$$

$$\sum_{i=1}^{3} d_i = 16 \leq 18 = 3(3-1) + (3+3+3+3);$$

$$\sum_{i=1}^{4} d_i = 21 \leq 23 = 4(4-1) + (4+4+3);$$

$$\sum_{i=1}^{5} d_i = 25 \leq 27 = 5(5-1) + (4+3);$$

$$\sum_{i=1}^{6} d_i = 29 \leq 33 = 6(6-1) + (3);$$

It satisfies all the inequalities and thus is graphical. However, if we consider 7, 6, 5, 5, 4, 4, 3,

$$\sum_{i=1}^{1} d_i = 7 \le 6 = 1(1-1) + (1+1+1+1+1+1);$$

this sequence fails the first inequality and thus is not graphical.

Split graphs are characterized as satisfying the mth Erdos and Gallai inequality by equality [Hammer, 1981], where m represents the largest index i in a degree sequence such that $d_i \ge i-1$. If this is the case, then the m nodes of largest degree form a clique in G, and the remaining nodes constitute the independent set.

We will refer to a split graph as *ideal* if every node in the clique that is adjacent to cones is adjacent to the same number of cones, and *proper* if all cones have the same degree. The notation for a split graph which is both ideal and proper will be *x-IPS(c,d,b)*, where x is the number of cone nodes to which each clique node is adjacent, c is the number of cone nodes, d is the common degree of each cone, and b is the number of clique nodes not adjacent to any cone. These parameters uniquely determine each of these split graphs; *x-IPS(c,d,b)* graphs exist for positive integers c, d, x and nonnegative integer b such that $\dfrac{cd}{x} \in \mathbb{N}$.

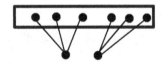

Figure 3.11: A split graph. We denote this graph 1-IPS(2,3,0), whose clique is of order 6, and observe that both sides of the 6th Erdos and Gallai inequality are equal to 36.

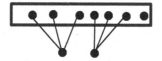

Figure 3.12: The split graph 1-IPS(2,3,1).

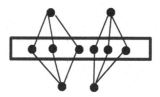

Figure 3.13: The split graph 2-IPS(4,3,0).

Theorem 1.5 can be applied to a variety split graphs in order to calculate their number of spanning trees. The following theorem summarizes those results for the number of spanning trees of some ideal proper non-threshold split graphs.

Theorem 3.14 *The number of spanning trees for some x-IPS(c,d,b) graphs are as follows:*

(1) *[Michewicz, 2006] If x = 1 and b = 0, then t(1-IPS(c,d,0)) =*

$$\frac{1}{n}(d+1)(n-c+1)^{n-2c} X^{c-1}Y^{c-1}$$

where $X = \left(\dfrac{[d(c+1)+1]+\sqrt{(c-1)^2 d^2 + 2(c+1)d + 1]}}{2} \right)$,

and $Y = \left(\dfrac{[d(c+1)+1]-\sqrt{(c-1)^2 d^2 + 2(c+1)d + 1]}}{2} \right)$.

(Figure 3.11)

(2) *[Moore, 2013] If x = 1 and b ≠ 0, then t(1-IPS(c,d,b)) =*

$$\frac{1}{n}(cd+b)^{b-1}(cd+b+1)^{c(d-1)}W_1\cdot W_2\cdot Z_1\cdot Z_2, \text{ where } W_1 =$$

$$\frac{1}{2}\left((cd+b)+d+1+\sqrt{(cd+b)^2+2(cd+b)(d+1)+(d+1)^2-4dn}\right),$$

$$W_2=\frac{1}{2}\left((cd+b)+d+1-\sqrt{(cd+b)^2+2(cd+b)(d+1)+(d+1)^2-4dn}\right),$$

$$Z_1=\frac{1}{2}\left((cd+b)+d+1+\sqrt{(cd+b)^2-2(cd+b)(d-1)+(d+1)^2}\right),$$

$$\text{and } Z_2=\frac{1}{2}\left((cd+b)+d+1-\sqrt{(cd+b)^2-2(cd+b)(d-1)+(d+1)^2}\right).$$

(*Figure* 3.12)

(3) *[Fuller, 2014] If* $x >1$ *and* $b = 0$, *then* $t(x\text{-}IPS(c,d,0)) =$

$$\frac{1}{c+\dfrac{cd}{x}}(x+d)\left(x+\frac{cd}{x}\right)^{\frac{c}{x}(d-1)}d^{c\left(1-\frac{1}{x}\right)}(P+Q)^{\frac{c}{x}-1}(P-Q)^{\frac{c}{x}-1},$$

where $P=\dfrac{1}{2}\left(x+d+\dfrac{cd}{x}\right)$, *and* $Q=\dfrac{1}{2}\sqrt{\left(x+d+\dfrac{cd}{x}\right)^2-4\dfrac{cd^2}{x}}.$

(*Figure* 3.13)

Proof Parts (1) and (2) of Theorem 3.14 were proven in the referenced publications, while part (3) has been submitted for publication. All of the proofs follow from elementary methods involving matrix row operations. To give a flavor for these proofs, we provide a sketch of the proof of (2). Consider the Temperley's B-Matrix, **B**, as defined in Chapter 1. Based on the structure of the $x\text{-}IPS(c, d, b)$ graphs when $b > 0$, we choose to arrange the matrix such that the nodes representing each row (column) are in decreasing order of degree. Hence, $\mathbf{B}-\lambda\mathbf{I}$ is of the

form $\begin{bmatrix}\mathbf{Q_1}&\mathbf{Q_4}\\\mathbf{Q_2}&\mathbf{Q_3}\end{bmatrix}$. Submatrix $\mathbf{Q_1}$ is a diagonal $(cd+b)\times(cd+b)$

matrix having $cd+b+1-\lambda$ as its first cd entries and $cd+b-\lambda$ as its remaining b entries, i.e.,

$$\mathbf{Q}_1 = \begin{bmatrix} cd+b+1-\lambda & 0 & 0 & & \cdots & & 0 & 0 \\ 0 & & \ddots & 0 & & \cdots & & 0 & 0 \\ 0 & & 0 & cd+b+1-\lambda & 0 & & & \vdots & \vdots \\ \vdots & & \vdots & 0 & & cd+b-\lambda & 0 & \vdots \\ 0 & & 0 & \cdots & & 0 & & \ddots & 0 \\ 0 & & \cdots & \cdots & & \cdots & & 0 & cd+b-\lambda \end{bmatrix}.$$

Submatrix $\mathbf{Q}_2$ is $c\times(cd+b)$. Since each cone node is adjacent to d clique nodes, we have the lefthand part of $\mathbf{Q}_2$ is a cascading vector of d zeros with the remainder of the entries as ones, and the righthand part of $\mathbf{Q}_2$ is a $c\times b$ matrix of ones, i.e.,

$$\mathbf{Q}_2 = \begin{bmatrix} 0\,0\,\cdots 0 & 1\,1\,\cdots 1 & \cdots & 1\,1\,\cdots 1 & 1 & \cdots & 1 \\ 1 & 0\,0\,\cdots 0 & 1\,1\,\cdots 1 & 1\,1\,\cdots 1 & \vdots & 1 & \vdots \\ \vdots & 1 & \ddots & 1\,1\,\cdots 1 & 1 & \vdots & 1 \\ 1 & \cdots & 1 & 0\,0\,\cdots 0 & 1\,1 & 1 \end{bmatrix}.$$

The bottom right quadrant, which we have labeled submatrix $\mathbf{Q}_3$, is a $c\times c$ matrix having $d+1-\lambda$ as each entry on its main diagonal and all ones off its main diagonal, i.e.,

$$\mathbf{Q}_3 = \begin{bmatrix} d+1-\lambda & 1 & \cdots & 1 \\ 1 & d+1-\lambda & 1 & \vdots \\ \vdots & 1 & \ddots & 1 \\ 1 & 1 & 1 & d+1-\lambda \end{bmatrix}.$$

Finally, $\mathbf{Q}_4 = \mathbf{Q}_2^T$.

We create an upper triangular matrix using row operations in an algorithmic fashion. First, we use a two-stage process to transform $\mathbf{Q}_2$ into a submatrix consisting of all zeros. We zero out the left-hand $c \times (cd)$ entries, corresponding to the clique nodes which are adjacent to cones, by multiplying the rows of $\mathbf{Q}_1$ that have diagonal entry $cd + b + 1 - \lambda$ by $-\dfrac{1}{cd + b + 1 - \lambda}$ and adding to the appropriate rows of $\mathbf{Q}_2$. Then we proceed to the rightmost $c \times b$ part of $\mathbf{Q}_2$, multiplying the rows of $\mathbf{Q}_1$ that have diagonal entry $cd + b - \lambda$ by $-\dfrac{1}{cd + b - \lambda}$ and adding to the appropriate rows of $\mathbf{Q}_2$. After this is completed, the matrix has $\mathbf{Q}_1$ and $\mathbf{Q}_4$ unchanged, the matrix $\mathbf{Q}_2$ is all zeros, and $\mathbf{Q}_3$ has all main diagonal entries the same, say, A, and all off- diagonal the terms the same, say, B, where

$$A = (d + 1 - \lambda) - \frac{d(c-1)}{(cd + b + 1) - \lambda} - \frac{b}{(cd + b) - \lambda}, \text{ and}$$

$$B = 1 - \frac{d(c-2)}{(cd + b + 1) - \lambda} - \frac{b}{(cd + b) - \lambda}.$$

Since $\mathbf{Q}_3 = \begin{bmatrix} A & B & B & \cdots & B \\ B & A & B & \cdots & B \\ B & B & \ddots & B & B \\ \vdots & \vdots & B & A & B \\ B & B & \cdots & B & A \end{bmatrix}$, the algorithmic application of row

operations transforms $\mathbf{Q}_3$ into an upper triangular submatrix, which means that the whole matrix is now upper triangular. Multiplying down the main diagonal, we obtain the product

$$(cd+b+1-\lambda)^{cd}(cd+b-\lambda)^{b}(A)\left(\frac{(A+B)(A-B)}{A}\right)\left(\frac{(A+2B)(A-B)}{(A+B)}\right)$$

$$\left(\frac{(A+3B)(A-B)}{(A+2B)}\right)\cdots\left(\frac{(A+(c-2)B)(A-B)}{(A+(c-3)B)}\right)\left(\frac{(A+(c-1)B)(A-B)}{(A+(c-2)B)}\right).$$

After cancellation, we have

$$(cd+b+1-\lambda)^{c(d-1)}(cd+b-\lambda)^{b-1}\left(\lambda^2-(cd+b+1)\lambda+d(cd+b)\right)^{c-1}$$
$$\left(\lambda^2-(cd+b+d+1)\lambda+d(cd+b+c)\right)(cd+b+c-\lambda).$$

The roots of this polynomial are multiplied together, after dividing out the n^2 (as in Theorem 1.13), $t(1\text{-}IPS(c,d,b))=$

$$\frac{1}{n}\times(cd+b)^{b-1}\times(cd+b+1)^{c(d-1)}\times W_1\times W_2\times Z_1\times Z_2,$$

where $W_1=\frac{1}{2}\left((cd+b)+d+1+\sqrt{(cd+b)^2+2(cd+b)(d+1)+(d+1)^2-4dn}\right)$,

$W_2=\frac{1}{2}\left((cd+b)+d+1-\sqrt{(cd+b)^2+2(cd+b)(d+1)+(d+1)^2-4dn}\right)$,

$Z_1=\frac{1}{2}\left((cd+b)+d+1+\sqrt{(cd+b)^2-2(cd+b)(d-1)+(d+1)^2}\right)$,

and $Z_2=\frac{1}{2}\left((cd+b)+d+1-\sqrt{(cd+b)^2-2(cd+b)(d-1)+(d+1)^2}\right)$. $\qquad\square$

We demonstrate the formulas presented in Theorem 3.14 in our next example.

Example 3.5 a Spanning trees for 1-IPS(2,3,0) (Figure 3.11). By Theorem 3.14(1), $t(1\text{-}IPS(c,d,0))=$

$$\frac{1}{n}(d+1)(n-c+1)^{n-2c} \cdot$$

$$\left(\frac{[d(c+1)+1]+\sqrt{(c-1)^2 d^2 + 2(c+1)d+1]}}{2}\right)^{c-1} \cdot$$

$$\left(\frac{[d(c+1)+1]-\sqrt{(c-1)^2 d^2 + 2(c+1)d+1]}}{2}\right)^{c-1} \cdot$$

For 1-IPS(2,3,0) , $c = 2$, $d = 3$ and $n = 8$, so

$$t\left(1-IPS(2,3,0)\right) = \frac{1}{8}(3+1)(8-2+1)^{8-2(2)} \cdot$$

$$\left(\frac{[3(2+1)+1]+\sqrt{(2-1)^2 3^2 + 2(2+1)3+1]}}{2}\right)^{2-1} \cdot$$

$$\left(\frac{[3(2+1)+1]-\sqrt{(2-1)^2 3^2 + 2(2+1)3+1]}}{2}\right)^{2-1} = 21609.$$

One Laplacian matrix for 1-IPS(2,3,0) is

$$\begin{bmatrix}
6 & -1 & -1 & -1 & -1 & -1 & -1 & 0 \\
-1 & 6 & -1 & -1 & -1 & -1 & -1 & 0 \\
-1 & -1 & 6 & -1 & -1 & -1 & -1 & 0 \\
-1 & -1 & -1 & 6 & -1 & -1 & 0 & -1 \\
-1 & -1 & -1 & -1 & 6 & -1 & 0 & -1 \\
-1 & -1 & -1 & -1 & -1 & 6 & 0 & -1 \\
-1 & -1 & -1 & 0 & 0 & 0 & 3 & 0 \\
0 & 0 & 0 & -1 & -1 & -1 & 0 & 3
\end{bmatrix}$$

which has (1,1)-cofactor 21609 as well, confirming the result of the formula.

Example 3.5 b Spanning trees for 1-IPS(2,3,0) (Figure 3.12).

By Theorem 3.14(2),

$$t\left(1-IPS\left(c,d,b\right)\right)=\frac{1}{n}\left(cd+b\right)^{b-1}\left(cd+b+1\right)^{c(d-1)}\cdot$$

$$\frac{1}{2}\left(\left(cd+b\right)+d+1+\sqrt{\left(cd+b\right)^2+2\left(cd+b\right)\left(d+1\right)+\left(d+1\right)^2-4dn}\right)\cdot$$

$$\frac{1}{2}\left(\left(cd+b\right)+d+1-\sqrt{\left(cd+b\right)^2+2\left(cd+b\right)\left(d+1\right)+\left(d+1\right)^2-4dn}\right)\cdot$$

$$\frac{1}{2}\left(\left(cd+b\right)+d+1+\sqrt{\left(cd+b\right)^2-2\left(cd+b\right)\left(d-1\right)+\left(d+1\right)^2}\right)\cdot$$

$$\frac{1}{2}\left(\left(cd+b\right)+d+1-\sqrt{\left(cd+b\right)^2-2\left(cd+b\right)\left(d-1\right)+\left(d+1\right)^2}\right).$$

For 1-IPS(2,3,1), $c = 2$, $d = 3$, $b = 1$ and $n = 9$, so

$$t\left(1-IPS(2,3,1)\right)=\frac{1}{9}\left(\left(2\right)\left(3\right)+1\right)^{1-1}\left(\left(2\right)\left(3\right)+1+1\right)^{2(3-1)}\cdot$$

$$\frac{1}{2}\left(11+\sqrt{\left(\left(2\right)\left(3\right)+1\right)^2+2\left(\left(2\right)\left(3\right)+1\right)\left(3+1\right)+\left(3+1\right)^2-4\left(3\right)\left(9\right)}\right)\cdot$$

$$\frac{1}{2}\left(11-\sqrt{\left(\left(2\right)\left(3\right)+1\right)^2+2\left(\left(2\right)\left(3\right)+1\right)\left(3+1\right)+\left(3+1\right)^2-4\left(3\right)\left(9\right)}\right)\cdot$$

$$\frac{1}{2}\left(11+\sqrt{\left(\left(2\right)\left(3\right)+1\right)^2-2\left(\left(2\right)\left(3\right)+1\right)\left(3-1\right)+\left(3+1\right)^2}\right)\cdot$$

$$\frac{1}{2}\left(11-\sqrt{\left(\left(2\right)\left(3\right)+1\right)^2-2\left(\left(2\right)\left(3\right)+1\right)\left(3-1\right)+\left(3+1\right)^2}\right)=$$

$$\frac{1}{9}\cdot8^4\cdot\frac{11+\sqrt{13}}{2}\cdot\frac{11-\sqrt{13}}{2}\cdot\frac{11+\sqrt{37}}{2}\cdot\frac{11-\sqrt{37}}{2}=258048.$$

One Laplacian matrix for 1-IPS(2,3,1) is

$$\begin{bmatrix}
7 & -1 & -1 & -1 & -1 & -1 & -1 & -1 & 0 \\
-1 & 7 & -1 & -1 & -1 & -1 & -1 & -1 & 0 \\
-1 & -1 & 7 & -1 & -1 & -1 & -1 & -1 & 0 \\
-1 & -1 & -1 & 7 & -1 & -1 & -1 & 0 & -1 \\
-1 & -1 & -1 & -1 & 7 & -1 & -1 & 0 & -1 \\
-1 & -1 & -1 & -1 & -1 & 7 & -1 & 0 & -1 \\
-1 & -1 & -1 & -1 & -1 & -1 & 6 & 0 & 0 \\
-1 & -1 & -1 & 0 & 0 & 0 & 0 & 3 & 0 \\
0 & 0 & 0 & -1 & -1 & -1 & 0 & 0 & 3
\end{bmatrix}$$

which has (1,1)-cofactor 258048 as well, confirming the result of the formula.

Example 3.5 c Spanning trees for 2-IPS(4,3,0) (Figure 3.13). By Theorem 3.14(3), $t(x\text{-}IPS(c,d,0)) =$

$$\frac{1}{c+\dfrac{cd}{x}}(x+d)\left(x+\frac{cd}{x}\right)^{\frac{c}{x}(d-1)}d^{c\left(1-\frac{1}{x}\right)} \cdot$$

$$\left(\frac{1}{2}\left(x+d+\frac{cd}{x}\right)+\frac{1}{2}\sqrt{\left(x+d+\frac{cd}{x}\right)^2-4\frac{cd^2}{x}}\right)^{\frac{c}{x}-1} \cdot$$

$$\left(\frac{1}{2}\left(x+d+\frac{cd}{x}\right)-\frac{1}{2}\sqrt{\left(x+d+\frac{cd}{x}\right)^2-4\frac{cd^2}{x}}\right)^{\frac{c}{x}-1}.$$

For 2-IPS(4,3,0), $x = 2$, $c = 4$, $d = 3$, $b = 0$, and $n = 10$, so $t(2\text{-}IPS(4,3,0)) =$

$$\frac{1}{4+\dfrac{(4)(3)}{2}}\cdot(2+3)\left(2+\dfrac{(4)(3)}{2}\right)^{\frac{4}{2}(3-1)}\cdot 3^{4\left(1-\frac{1}{2}\right)}\bullet$$

$$\left(\frac{1}{2}\left(2+3+\dfrac{(4)(3)}{2}\right)+\frac{1}{2}\sqrt{\left(2+3+\dfrac{(4)(3)}{2}\right)^2-4\dfrac{(4)(3)^2}{2}}\right)^{\frac{4}{2}-1}\bullet$$

$$\left(\frac{1}{2}\left(2+3+\dfrac{(4)(3)}{2}\right)-\frac{1}{2}\sqrt{\left(2+3+\dfrac{(4)(3)}{2}\right)^2-4\dfrac{(4)(3)^2}{2}}\right)^{\frac{4}{2}-1}=$$

$$331776.$$

One Laplacian matrix for 2-IPS(4,3,0) is

$$\begin{bmatrix}
7 & -1 & -1 & -1 & -1 & -1 & -1 & -1 & 0 & 0 \\
-1 & 7 & -1 & -1 & -1 & -1 & -1 & -1 & 0 & 0 \\
-1 & -1 & 7 & -1 & -1 & -1 & -1 & -1 & 0 & 0 \\
-1 & -1 & -1 & 7 & -1 & -1 & 0 & 0 & -1 & -1 \\
-1 & -1 & -1 & -1 & 7 & -1 & 0 & 0 & -1 & -1 \\
-1 & -1 & -1 & -1 & -1 & 7 & 0 & 0 & -1 & -1 \\
-1 & -1 & -1 & 0 & 0 & 0 & 3 & 0 & 0 & 0 \\
-1 & -1 & -1 & 0 & 0 & 0 & 0 & 3 & 0 & 0 \\
0 & 0 & 0 & -1 & -1 & -1 & 0 & 0 & 3 & 0 \\
0 & 0 & 0 & -1 & -1 & -1 & 0 & 0 & 0 & 3
\end{bmatrix}$$

which has (1,1)-cofactor 331776 as well, confirming the result of the formula.

Chapter 4

Approaches to the Multigraph Problem

When optimizing the number of spanning trees over a class of graphs, the problem is fairly straightforward. If we consider the condition that each pair of nodes has at most one edge between them, and then "relax" that condition to allow for multigraphs (i.e., allow for multiple edges between a pair of nodes), one would surmise that the solution to the problem would be readily attainable. That is simply not the case for the spanning tree problem. In this chapter, we explore some possible approaches to one problem in particular.

Kelmans and Chelnokov [Kelmans, 1974] and Shier [Shier, 1974] independently proved that the graph $K_n - kK_2$, $k \leq \dfrac{n}{2}$ has the greatest number of spanning trees among all graphs in its class $\Omega\left(n, \dbinom{n}{2} - k\right)$. We conjecture that this graph also maximizes the number of spanning trees over all multigraphs in the class having at most a single multiple edge of multiplicity equal to two. Based on additional evidence, we conjecture that $K_n - kK_2$ maximizes the number of spanning trees over all multigraphs in the class.

147

We define a *full neighborhood point (fnp)* to be a node w such that $N(w) = V - \{w\}$. If we consider the case of a connected multigraph with exactly one double edge, we consider the placement of that multiple edge in relation to any fnp.

The end nodes of the lone double edge can exist in any of the following situations:

1. exactly one end node is a fnp;
2. both end nodes are fnp's;
3. neither end node is a fnp.

We denote by G_s the underlying graph of the multigraph M.

Case 1: The multiple edge joins a fnp w and a point v which is not a fnp.

Since $N(v) \neq V - \{v\}$, there exists a node $y \notin N(v)$. Thus, the neighborhood of w dominates that of y, and, applying Theorem 3.8 (i.e., the Swing Surgery), one of the two edges $\{w,v\}$ can be replaced by $\{v,y\}$. The number of spanning trees has not decreased; the multigraph has been transformed into a graph G'; and $t(n,M) \leq t(n,G') \leq t(K_n - kK_2)$. If $G' \neq K_n - kK_2$, the second inequality is strict. If $G' = K_n - kK_2$ then $G_s = K_n - (k+1)K_2$ and exactly one end node of the multiple edge is the end node of the deleted matching edge. Let x be the edge $\{w,v\}$ in M that was removed and replaced by $x' = \{v,y\}$ from G'. Observe that $M - x = G' - x' = K_n - (k+1)K_2$ and

$G' \mid x'$ is obtained from the connected graph $G \mid x$ (with loop deleted) by the addition of an edge. Thus, $t(M \mid x) < t(G' \mid x')$ and $t(M) < t(G')$. Hence, $t(M) < t(K_n - kK_2)$.

Case 2: The multiple edge joins two fnp's.

Suppose the multiple edge joins w and v, both of which are of full neighborhood. Because $e = \binom{n}{2} - k$, there must exist at least two nodes u and u' that are not adjacent. Clearly, $N(v)$ dominates $N(u')$, so again employing the Swing Surgery, edge $\{v,u\}$ can be replaced by edge $\{u,u'\}$ without reducing the number of spanning trees. Now w and v satisfy the condition of case (1) in the new graph, and so again we have $t(M) < t(K_n - kK_2)$.

Case 3: The multiple edge is incident on two nodes each not of full neighborhood.

Case 3 is much more involved, but is solved for many instances of multigraphs in the class. To get a sense of the problem, we require some further algebraic development. Given two vectors of nonnegative integers $\mathbf{x}$ and $\mathbf{y}$, both of which are of length n, and whose elements are nonincreasing, $\mathbf{x}$ *is majorized by* $\mathbf{y}$, denoted $\mathbf{xx} \prec \mathbf{y}$, if $\sum_{i=1}^{k} x_i \leq \sum_{i=1}^{k} y_i$ for each $1 \leq k \leq n-1$ and $\sum_{i=1}^{n} x_i = \sum_{i=1}^{n} y_i$ [Marshall, 1979]. In other words, to establish majorization, the direction of the inequalities for the partial sums cannot change until the final entries in the respective vectors.

Majorization gives an indication of how "spread out" the elements of a vector are. The more spread out, the smaller the product of the elements, thus, $\mathbf{xx} \prec \mathbf{y}$ implies $\prod_{i=1}^{n} x_i \geq \prod_{i=1}^{n} y_i$ [Marshall 1979].

Example 4.1 Consider the following three vectors $\mathbf{x} = (15,10,10,5,3,3)$, $\mathbf{y} = (15,12,11,4,3,1)$ and $\mathbf{z} = (16,12,7,4,4,3)$. The table below lists the partial sums.

$\sum_{i=1}^{k}$	$\mathbf{x} = (15,10,10,5,3,3)$	$\mathbf{y} = (15,12,11,4,3,1)$	$\mathbf{z} = (16,12,7,4,4,3)$
$k = 1$	15	15	16
$k = 2$	25	27	28
$k = 3$	35	38	35
$k = 4$	40	42	39
$k = 5$	43	45	43
$k = 6$	46	46	46

Clearly, $\mathbf{x} \prec \mathbf{y}$, as $\sum_{i=1}^{k} x_i \leq \sum_{i=1}^{k} y_i$ for all $k \leq 5$, and $\sum_{i=1}^{6} x_i = \sum_{i=1}^{6} y_i$. As indicated $\prod_{i=1}^{6} x_i = 67500 \geq 23760 = \prod_{i=1}^{6} y_i$. We note that there is no further majorization relationship among the three vectors. For example, $\mathbf{x} \nprec \mathbf{z}$ and $\mathbf{z} \nprec \mathbf{x}$, as

$$\sum_{i=1}^{2} x_i = 25 < 28 = \sum_{i=1}^{2} z_i \text{ , but } \sum_{i=1}^{4} x_i = 40 > 39 = \sum_{i=1}^{4} z_i \text{ .}$$

Similarly, neither **x** nor **z** majorize each other.

To prove our next result, namely, for a positive definite symmetric matrix **L**, $\lambda(\mathbf{L}) \succ \mathbf{d}(\mathbf{L})$ i.e., the vector of the diagonals of **L** is majorized by the vector of its eigenvalues [Marshall, 1979], we require some further development. Given two non-increasing sequences of real numbers $\mathbf{d} = (d_n, d_{n-1}, ..., d_1)$ and $\mathbf{d}' = (d'_m, d'_{m-1}, ..., d'_1)$, $n > m$, **d'** is said to *interlace* **d** if $d_i \le d'_i \le d_{n-m+i}$, $i = 1, ..., m$.

Example 4.2 Let $\mathbf{d} = (42, 32, 22, 12, 2)$ and $\mathbf{d}' = (25, 15, 5)$, thus $n = 5$, $m = 3$ and $n - m + i = 2 + i$. We note that $d_1 \le d'_1 \le d_3$ $(2 \le 5 \le 22)$, $d_2 \le d'_2 \le d_4$ $(12 \le 15 \le 32)$, and $d_3 \le d'_3 \le d_5$ $(22 \le 25 \le 42)$, so **d'** interlaces **d**, $(42, 32, 25, 22, 15, 12, 5, 2)$.

We use this notion in our next lemma, which is sometimes attributed to Cauchy.

Lemma 4.1 *Let* **A** *be a symmetric positive definite* $(n-1) \times (n-1)$ *matrix,* **y** *an* $(n-1)$*-dimensional vector, and* $a \in \mathbb{R}$*. Denote by* $\hat{\mathbf{A}}$ *the* $n \times n$ *matrix* $\hat{\mathbf{A}} = \begin{bmatrix} \mathbf{A} & \mathbf{y} \\ \mathbf{y}^T & a \end{bmatrix}$*. In addition, let* $\lambda_1 \le \lambda_2 \le ... \le \lambda_{n-1}$ *represent the eigenvalues of* **A** *and* $\hat{\lambda}_1 \le \hat{\lambda}_2 \le ... \le \hat{\lambda}_n$ *represent the eigenvalues of*

$\hat{\mathbf{A}}$. *Then* $\hat{\lambda}_1 \leq \lambda_1 \leq \hat{\lambda}_2 \leq \lambda_2 ... \leq \hat{\lambda}_{n-1} \leq \lambda_{n-1} \leq \hat{\lambda}_n$, *i.e., the eigenvalues of* $\mathbf{A}$ *interlace those of* $\hat{\mathbf{A}}$.

Proof It suffices to show that $\hat{\lambda}_j \leq \lambda_j \leq \hat{\lambda}_{j+1}$. Let $\mathbf{x} \in \mathbb{R}^{n-1}$, $z \in \mathbb{R}$, L_j an arbitrary $(n-j+1)$-dimensional subspace *of* $\mathbb{R}^n$ and $\hat{\mathbf{x}} = \begin{bmatrix} \mathbf{x} \\ z \end{bmatrix}$. By Theorem 0.5 (Courant-Fischer),

$$\hat{\lambda}_j = \max_{L_j} \min_{0 \neq \mathbf{x} \in L_j} \frac{\hat{\mathbf{x}}^T \hat{\mathbf{A}} \hat{\mathbf{x}}}{\hat{\mathbf{x}}^T \hat{\mathbf{x}}} \leq \max_{L_{n-j+1}} \min_{0 \neq \mathbf{x} \in L_{n-j+1}} \frac{\hat{\mathbf{x}}^T \hat{\mathbf{A}} \hat{\mathbf{x}}}{\hat{\mathbf{x}}^T \hat{\mathbf{x}}} = \max_{L_{n-j+1}} \min_{0 \neq \mathbf{x} \in L_{n-j+1}} \frac{\mathbf{x}^T \mathbf{A} \mathbf{x}}{\mathbf{x}^T \mathbf{x}} = \lambda_{j-1}.$$

The upper bound $\lambda_j \leq \hat{\lambda}_{j+1}$ can be demonstrated similarly.   □

We use this lemma to prove the theorem:

Theorem 4.2 *Let* $\mathbf{A}$ *be a symmetric positive definite* $n \times n$ *matrix, with diagonal entries arranged in non-decreasing order. Then the vector of diagonals of* $\mathbf{A}$, $\mathbf{d} = (d_n, d_{n-1}, ..., d_1)$, *majorizes the vector of eigenvalues of* $\mathbf{A}$, $\lambda = (\lambda_n, \lambda_{n-1}, ..., \lambda_1)$, *i.e.,* $\mathbf{d} \succ \lambda$.

Proof By induction on n. Assume that the result is true for all $j \times j$ matrices satisfying the hypotheses, $2 \leq j \leq n-1$. Let $\mathbf{A}$ be an $n \times n$ matrix and $\mathbf{A}_n$ be a submatrix of $\mathbf{A}$ obtained by deleting the nth row and n^{th} column. Let $\lambda_1 \leq \lambda_2 \leq ... \leq \lambda_n$ be eigenvalues of $\mathbf{A}$ and $\hat{\lambda}_1 \leq \hat{\lambda}_2 \leq ... \leq \hat{\lambda}_{n-1}$ the eigenvalues of $\mathbf{A}_n$. By the induction hypothesis, we have $\displaystyle\sum_{i=1}^{j} d_{n-i} \geq \sum_{i=1}^{j} \hat{\lambda}_{n-i}$, $j = 1, \ldots, n-1$. By Lemma 4.6, we have

$$\lambda_1 \leq \hat{\lambda}_1 \leq \lambda_2 \leq \hat{\lambda}_2 \ldots \leq \lambda_{n-1} \leq \hat{\lambda}_{n-1} \leq \lambda_n, \text{ so } \sum_{i=1}^{j} \hat{\lambda}_{n-i} \geq \sum_{i=1}^{j} \lambda_{n-i}, \; j = 1, \ldots, n-1.$$

Therefore, $\displaystyle\sum_{i=1}^{j} d_{n-i+1} \geq \sum_{i=1}^{j} \lambda_{n-i+1}$, $j = 1, \ldots, n-1$, and, by Theorem 0.11,

have $\displaystyle\sum_{i=1}^{n} d_i = \sum_{i=1}^{n} \lambda_i$, i.e., the trace is the sum of the eigenvalues. □

Now consider the **B**-matrix of $K_n - kK_2$ and the **B**-matrix of a multigraph M in the same class. Since the **B**-matrix is positive definite and symmetric,

$$\prod_{i=1}^{n} d_i(\mathbf{B}(M)) \geq \prod_{i=1}^{n} \lambda_i(\mathbf{B}(M)) \text{ and}$$

$$\prod_{i=1}^{n} d_i\left(\mathbf{B}\left(K_n - kK_2\right)\right) \geq \prod_{i=1}^{n} \lambda_i\left(\mathbf{B}\left(K_n - kK_2\right)\right).$$

Let M be such that

$$\mathbf{d}\left(\mathbf{B}(M)\right) \succ \lambda\left(\mathbf{B}\left(K_n - kK_2\right)\right).$$

Then

$$\prod_{i=1}^{n} d_i(\mathbf{B}(M)) \leq \prod_{i=1}^{n} \lambda_i(\mathbf{B}(K_n - kK_2))$$

which in turn implies

$$\prod_{i=1}^{n} \lambda_i(\mathbf{B}(M)) \leq \prod_{i=1}^{n} \lambda_i(\mathbf{B}(K_n - kK)),$$

and thus such a multigraph has fewer spanning trees. Using majorization, many multigraphs can be shown to have fewer spanning trees than $K_n - kK_2$ in their respective classes.

Recall that the **B**-matrix eigenvalue spectrum of $K_n - kK_2$ is $n - k$ instances of n and k instances of $n - 2$. In the overwhelming number of cases validated by computer, the diagonals of the **B**-matrix of these multigraphs majorize the eigenvalues of the **B**-matrix of the corresponding $K_n - kK_2$ in its class, and therefore, the multigraphs have fewer spanning trees. Hence, aside from a relatively small number of exceptional cases, the majorization argument is valid. The exceptions can be classified by the following theorem:

Theorem 4.3 *The vector of diagonals of the **B**-matrix of a multigraph M will not majorize the vector of eigenvalues of a $K_n - kK_2$ in the same class if and only if*

$$\sum_{i=n-k+1}^{n} d_i\left(\mathbf{B}(M)\right) > k(n-2) = \sum_{i=n-k+1}^{n} \lambda_i\left(\mathbf{B}(K_n - kK_2)\right).$$

*i.e., the sum of the last k elements of the vector of diagonals of the multigraph's **B**-matrix is greater than the sum of the corresponding elements of the vector of $K_n - kK_2$ **B**-matrix eigenvalues.*

Proof To simplify notation, we let d_i refer to the diagonals of the **B**-matrix of the multigraph and let λ_i refer to the eigenvalues for the unique $K_n - kK_2$ in its class. If $\sum_{i=n-k+1}^{n} d_i > k(n-2) = \sum_{i=n-k+1}^{n} \lambda_i$, then

$$\sum_{i=1}^{n} d_i - \sum_{i=n-k+1}^{n} d_i = \sum_{i=1}^{n-k} d_i < \sum_{i=1}^{n-k} \lambda_i = \sum_{i=1}^{n} \lambda_i - \sum_{i=n-k+1}^{n} \lambda_i \,.$$

Hence, the vector of multigraph **B**-matrix diagonals cannot majorize the vector of eigenvalues. Now, assume the vector of diagonals of a multigraph's **B**-matrix does not majorize the eigenvalues of $K_n - kK_2$.

Then there exists a $1 \le j < n$, so that $\sum_{i=1}^{j} d_i < \sum_{i=1}^{j} \lambda_i$. Assume j is the

smallest such index. Then $d_j < \lambda_j$, or else $\sum_{i=1}^{j-1} d_i < \sum_{i=1}^{j-1} \lambda_i$, which would

contradict the selection of j. Also, $j \le n - k$. If not, then

$d_j < \lambda_j = n - 2$, and since the d_j are nonincreasing, then $\sum_{i=j+1}^{n} d_i < \sum_{i=j+1}^{n} \lambda_i$,

and so,

$$\sum_{i=1}^{n} d_i = \sum_{i=1}^{j} d_i + \sum_{i=j+1}^{n} d_i < \sum_{i=1}^{j} \lambda_i + \sum_{i=j+1}^{n} \lambda_i = \sum_{i=1}^{n} \lambda_i \,,$$

which contradicts the equality of the sums. Therefore, $j \le n - k$, which implies $d_j < n$. In fact, d_j must be $n - 1$, or once again the sums would not be equal. Now, having established that

$$\sum_{i=j+1}^{n} d_i > \sum_{i=j+1}^{n} \lambda_i \text{,with } \sum_{i=j+1}^{n-k} d_i < \sum_{i=j+1}^{n-k} \lambda_i \,,$$

it follows that

$$\sum_{i=n-k+1}^{n} d_i < \sum_{i=n-k+1}^{n} \lambda_i = k(n-2)\,. \qquad \square$$

We prove the following corollary to the theorem.

Corollary 4.4 *If the vector of diagonals does not majorize the vector of eigenvalues, then* $d_1 \le 2n - k - 2$, *and* $d_n \ge n - k$.

Proof Given the non-majorization assumption, let j be the smallest index such that $\sum_{i=1}^{j} d_i < \sum_{i=1}^{j} \lambda_i$. We have shown that $j \le n - k$, and $d_j = n - 1$, so that

$$\sum_{i=1}^{j} d_i \ge d_1 + (j-1)(n-1) = d_1 + jn - j - n + 1 \text{ and } \sum_{i=1}^{j} \lambda_i = jn.$$

Therefore, since $\sum_{i=1}^{j} d_i < \sum_{i=1}^{j} \lambda_i$, we have

$$d_1 + jn - j - n + 1 < jn, \text{ or } d_1 < n + j - 1.$$

Now, since

$$j \le n - k, d_1 < n + (n - k) - 1 = 2n - k - 1.$$

Because there is non-majorization,

$$\sum_{i=n-k+1}^{n} d_i > k(n-2) = \sum_{i=n-k+1}^{n} \lambda_i, \text{ and } d_{n-k+1} \le n - 1.$$

Thus,

$$\sum_{i=n-k+1}^{n} d_i < (k-1)(n-1) + d_n \text{ so } (k-1)(n-1) + d_n > k(n-2)$$

implying $d_n > n - k - 1$.   □

Even though majorization handles the vast majority of instances, the exceptional cases must be handled individually, and this cannot be done efficiently given the multitude of multigraph realizations for a given degree sequence. A specific example for which majorization will not hold is the **B**-matrix diagonal sequence with two distinct entries, n and $n-1$, where n occurs $n-2k$ times, and $n-1$ occurs $2k$ times. The eigenvalues for $\mathbf{B}\left(K_n - kK_2\right)$ are $\left(n,...,n,n-2,...,n-2\right)$ where n occurs with multiplicity $n-k$ and $n-2$ occurs with multiplicity k. The sum of the last k diagonals is $k\left(n-1\right)$ which is clearly greater than $k\left(n-2\right)$, so by Theorem 4.12, the vector of eigenvalues for the **B**-matrix is not majorized. Note that the multigraph specified above has the same degree sequence as $K_n - kK_2$ in its class. Hence, since the majorization is in the other direction, it cannot be applied here.

Chapter 5

Laplacian Integral Graphs and Multigraphs

Of particular interest is the classification of graphs by the nature of their eigenvalues. In this chapter, we discuss those graphs and multigraphs whose eigenvalues are all integers. There are many such graphs and multigraphs, and this chapter is not intended to be an exhaustive exposition. Much of this material has appeared in earlier chapters of this book.

5.1   Complete Graphs and Related Structures

The Laplacian eigenvalues for a complete graph K_n, as given in the proof of Cayley's Theorem (Proposition 1.6(a)) are 0, and $n-1$ instances of n. Of course, this leads to the fact that the eigenvalues for $\mathbf{H}(K_2)$ are 0 and 2. This fact can be used to prove the following proposition, which is used in Chapter 4.

Proposition 5.1   *The eigenvalues for* $\mathbf{H}(K_n - kK_2)$ *are* $n,...n,$ $n - 2,...,n - 2,0$, *where n occurs with multiplicity* $n - k - 1$ *and* $n - 2$ *occurs with multiplicity k.*

Proof Obviously, $k \leq \dfrac{n}{2}$. Now, $K_n - kK_2$ has as its complement k

copies of K_2, and if $k < \dfrac{n}{2}$, $n - 2k$ copies of K_1. Thus, the

eigenvalues of the complement of $K_n - kK_2$ are $n - k$ zeros, and k

instances of 2. By Proposition 1.4(a), if $\overline{G}$ is the complement of

graph G, and the eigenvalues of G are arranged in non-decreasing

order, then $\lambda_{k+2}(\overline{G}) = n - \lambda_{n-k}(G)$ for $k = 0, \dots, n-2$. Direct application

yields the result.   □

Remark 5.1 The proofs of the spanning tree formulas for complete
bipartite and regular complete k-partite graphs rely heavily on the fact
that the complements of these graphs are disjoint complete graphs. Thus,
these graphs are also Laplacian integral.

For the $K_n - P_p$ $(n > p)$ graphs in part (c)(i) of Proposition 1.12,

the complements are P_p, which do not have integer eigenvalues for

all values of p, so therefore, the graphs themselves do not have

integer eigenvalues, either. The same is true for $K_n - C_p$ graphs in

part (c)(ii); while C_3, C_4, and C_6 all have integer eigenvalues, this is

not the case for all cycles. For example, it can be shown by direct

application of Theorem 1.5 that C_8 has eigenvalue spectrum

$0, 2 - \sqrt{2}, 2 - \sqrt{2}, 2, 2, 2 + \sqrt{2}, 2 + \sqrt{2}, 4$, so its complement will not have

an integral spectrum, either.

5.2 Split Graphs and Related Structures

In Chapter 3, the Laplacian eigenvalues for various types of split graphs were introduced. Recall that a split graph is one in which the nodes can be partitioned into a clique and an independent set, while a threshold graph is a split graph with the neighborhood domination property, i.e, for all pairs of nodes u and v in G, $N(u) - \{v\} \subseteq N(v) - \{u\}$ whenever $\deg(u) \le \deg(v)$.

As Theorem 3.5 gives the Laplacian characteristic polynomial for threshold graphs in factored form, we observe that the eigenvalues are all integers. Until 2010, it was unknown if there existed Laplacian integral, non-threshold split graphs. Kirkland, Alverez de Freitas, Del Vecchio and de Abreu [Kirkland, 2010] proved the existence of an infinite such class of split graphs, the "biregular split graphs," using techniques such as Kronecker products, balanced incomplete block designs and solutions to certain Diophantine equations. The biregular split graph is the same graph as a certain subclass of the *x-IPS(c,d,0)* graphs presented in section 3 of Chapter 3, whose spanning tree formula can be proven using elementary matrix operations as outlined in the sketch of the proof of Theorem 3.14(2). The Laplacian integral nature of these graphs was discovered independently.

Recall that the spanning tree formula is given as

$$t(x - IPS(c,d,0) = \frac{1}{c + \dfrac{cd}{x}} (x + d) \left(x + \frac{cd}{x} \right)^{\frac{c}{x}(d-1)} d^{c\left(1 - \frac{1}{x}\right)} \bullet$$

$$\left(\frac{1}{2} \left(x + d + \frac{cd}{x} \right) + \frac{1}{2} \sqrt{\left(x + d + \frac{cd}{x} \right)^2 - 4 \frac{cd^2}{x}} \right)^{\frac{c}{x} - 1} \bullet$$

$$\left(\frac{1}{2}\left(x+d+\frac{cd}{x}\right)-\frac{1}{2}\sqrt{\left(x+d+\frac{cd}{x}\right)^2-4\frac{cd^2}{x}}\right)^{\frac{c}{x}-1}.$$

The characteristic polynomial giving the result is

$$\lambda(d-\lambda)^{c-\frac{c}{x}}\left(\lambda-(x+d)\right)\left(\frac{cd}{x}+x-\lambda\right)^{\frac{c}{x}(d-1)}\cdot$$

$$\left(\lambda^2-\left(x+d+\frac{cd}{x}\right)\lambda-\frac{cd^2}{x}\right)^{\frac{c}{x}-1},$$

which has roots

$$\lambda=0,(x+d),x+\frac{cd}{x}\text{ with multiplicity }\frac{c}{x}(d-1)\text{ , and }d,$$

$$\left(\frac{1}{2}\left(x+d+\frac{cd}{x}\right)+\frac{1}{2}\sqrt{\left(x+d+\frac{cd}{x}\right)^2-4\frac{cd^2}{x}}\right)$$

and

$$\left(\frac{1}{2}\left(x+d+\frac{cd}{x}\right)-\frac{1}{2}\sqrt{\left(x+d+\frac{cd}{x}\right)^2-4\frac{cd^2}{x}}\right)$$

each with multiplicity $c\left(1-\frac{1}{x}\right).$

We note that, if the discriminant is an odd perfect square, the graph will be Laplacian Integral. For example, the graph *3-IPS(6,2,0)* has Laplacian spectrum $0,2,2,2,2,5,7,7$, obtained by substituting $x=3$, $c=6$ and $d=2$. Some other examples of Laplacian integral *x-IPS(c,d,0)* graphs are *3-IPS(12,2,0)*, and *2-IPS(4,3,0)* . [Fuller, 2010]

There is another class of graphs, based on split graphs, whose eigenvalues are integral. A graph is defined as *almost split* if it is not a split graph, but can be transformed to one by the deletion of one edge. We will denote by AS(n,2) an almost split graph on n nodes, with a clique of order $n-2$, two adjacent nodes that are also adjacent to roughly half the clique nodes, with one clique node in their common neighborhood. The degrees of the non-clique nodes will be the same in the case where n is odd (i.e., their degrees will each be $\dfrac{n+1}{2}$), and if n is even, the degrees of these nodes will differ by one (i.e., one will be of degree $\dfrac{n}{2}$ and the other will be of degree $\dfrac{n}{2}+1$). When n is odd, AS(n,2) has degree sequence $\dfrac{n+1}{2}, \dfrac{n+1}{2}, \underbrace{n-2,...,n-2}_{n-3 \; terms}, n-1$. When n is even, AS(n,2) has degree sequence $\dfrac{n}{2}, \dfrac{n}{2}+1, \underbrace{n-2,...,n-2}_{n-3 \; terms}, n-1$. It is easy to show that such graphs will always have $\dfrac{n^2-3n+6}{2}$ edges, regardless of the parity of n. Figures 5.1 and 5.2 depict two examples of AS(n,2).

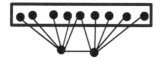

Figure 5.1: The graph AS(11,2).

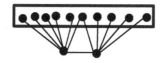

Figure 5.2: The graph AS(12,2).

Recall that a split graph satisfies the mth Erdos and Gallai inequality with equality [Hammer, 1981], where m represents the largest index i in a degree sequence such that $d_i \geq i - 1$. For the graph in Figure 5.1 (a), the degree sequence is 10, 9, 9, 9, 9, 9, 9, 9, 9, 6, 6; $m = 9$, and

$$\sum_{i=1}^{9} d_i = 82 \leq 84 = 9(9-1) + (6+6),$$ so the $m = 9^{th}$ Erdos and Gallai

inequality is not an equality. Therefore, AS(11,2) is not a split graph. This is the case for any AS(n, 2) graph.

The next proposition gives a formula for their eigenvalues.

Proposition 5.2 *(a) The eigenvalues for* $AS(n,2)$, n *odd, are*

$$0, \frac{n+1}{2}, \frac{n+1}{2}, \underbrace{n-1,...,n-1}_{n-5 \text{ terms}}, n, n,$$ *all of which are integers.*

(b) The eigenvalues for $AS(n,2)$, n *even, are* $0, \dfrac{n}{2}, \dfrac{n}{2}+1, \underbrace{n-1,...,n-1}_{n-5 \text{ terms}}, n, n,$

all of which are integers.

Proof The proof proceeds from the observation that the complement of AS(n,2) is comprised of three components: one K_1, and two disjoint stars.

(a) If n is odd, the complement of $AS(n,2)$ is one K_1, and two disjoint identical stars $K_{1,\frac{n-3}{2}}$. It can be deduced from

Proposition 1.12(b)(i) that a star $K_{1,\frac{n-3}{2}}$ has eigenvalues

$0,\ \underbrace{1,...,1}_{\frac{n-3}{2}-1\ terms},\ \dfrac{n-3}{2}+1.$ Thus, the eigenvalues for these three

components, arranged in non-decreasing order, are

$0,0,0,\ \underbrace{1,...,1}_{n-5\ terms},\dfrac{n-1}{2},\dfrac{n-1}{2},$ and so, by Proposition 1.4 (a),

$AS(n,2)$ has Laplacian spectrum

$$0,\dfrac{n+1}{2},\dfrac{n+1}{2},\underbrace{n-1,...,n-1}_{n-5\ terms},n,n.$$

(b) The proof for n even follows similarly. The only difference is
that, instead of two identical stars, the two stars in the
complement are $K_{1,\frac{n-2}{2}}$ and $K_{1,\frac{n-4}{2}}$.

Remark 5.2 The graph $AS(n,2)$, n even, has $\left(\dfrac{n+1}{2}\right)^2(n-1)^{n-5}n$

spanning trees, while $AS(n,2)$, n odd, has $\left(\dfrac{n}{2}\right)^2(n-1)^{n-5}n$ spanning

trees.

5.3 Laplacian Integral Multigraphs

There are several multigraphs that are shown to be Laplacian integral.
For example, by Theorem 1.30, gK_n (i.e., the complete graph with g
multiple edges between every pair of nodes) has eigenvalue zero, and
$n-1$ instances of gn.

Another Laplacian integral multigraph is one with an underlying threshold graph, first presented by Heinig and Saccoman [Heinig, 2012]. We describe the construction of such a graph as follows: we begin with a proper threshold graph on n nodes with n_1 nodes of degree v_1. Each edge of the induced complete subgraph induced by the common neighborhood of the cones is replicated μ times, and we denote this multigraph by $MT^{\mu}_{n;v_1^{n_1}}$. Figure 5.3 depicts $MT^3_{7;2^2}$.

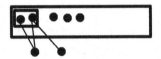

Figure 5.3: The multigraph $MT^3_{7;2^2}$ which has degree sequence 2, 2, 4, 4, 4, 8, 8.
The inner box in the clique houses a $3K_2$.

As in the case of threshold graphs, the eigenvalue spectrum for the MT graphs is predicated on the parity of the number of unique degree sequence terms.

Proposition 5.3 *[Heinig, 2012] The multigraph $MT^{\mu}_{n;v_1^{n_1}}$ has Laplacian eigenvalues as follows:*

    (a)   *(even case: $v_1 = n - n_1 - 1$)*

$$\lambda = (0, v_1^{n_1-1}, n, (v_2 + \mu)^{n_2-1}), \text{ degree sequence}$$

$$v_1^{n_1}, v_2^{n_2}.$$

(b)   *(odd case:* $v_1 < n - n_1 - 1$ *)*

$$\lambda = (0, v_1^{n_1}, (v_2+1)^{n_2-1}, n, (v_3+\mu)^{n_3-1}, \text{ degree sequence}$$

$$v_1^{n_1}, v_2^{n_2}, v_3^{n_3}.$$

The proof of Proposition 5.3 follows from a sequence of elementary matrix row operations, similar to the proof of Theorem 3.14(2).

Remark 5.3 The graph $MT^{\mu}_{n;v_1^{n_1}}$, in the even case, has

$v_1^{n_1-1}(v_2+1)^{n_2-1}(v_3+\mu)^{n_3-1}$ spanning trees, while $MT^{\mu}_{n;v_1^{n_1}}$, in the odd

case, has $v_1^{n_1-1}(v_2+\mu)^{n_2-1}$ spanning trees.

Bibliography

[Ball, 1982] M.O. Ball and S.J. Provan, *Bounds on the reliability polynomial for shellable independence systems,* SIAM J. Alg. Disc. Meth. 3, 166-181, 1982.

[Ball, 1983] M. Ball and J.S. Provan, *The complexity of counting cuts and of computing the probability that a graph is connected,* SIAM J. on Computing 12, 777-788,1983.

[Bapat, 2010] R. B. Bapat. *Graphs and Matrices.* Springer: London, Hindustan Book Agency, New Dehli, India, 2010.

[Bleiler, 2007] S. Bleiler, and J.T. Saccoman. *A Correction in the Formula for the Number of Spanning Trees in Threshold Graphs.* Australasian Journal of Combinatorics 37, 205-213, 2007.

[Bogdanowitz, 1985] Z. Bogdanowicz, *Spanning trees in undirected simple graphs,* Doctoral Dissertation, Stevens Tech, Computer Science Department, 1985.

[Bogdanowitz, 2009] Z. Bogdanowicz, *Undirected simple connected graphs with minimum number of spanning trees,* Discrete Math 309 (no. 10), 3074-3082, 2009.

[Boesch, 1984] F.T. Boesch and C.L. Suffel, *A survey of the algebraic approach to the study of spanning trees,* Technical Report, Stevens Tech, Computer Science Department, 8401, 1984.

[Brooks, 1940] R.L. Brooks, C.A.B. Smith, A.H. Stone and W.T. Tutte, *Dissection of a Rectangle into Squares,* Duke Mathematics Journal 7, 312-340, 1940.

[Brown, 1993]. J.J. Brown, C. Colbourn, and J.S. Devitt, *Network transformations and bounding network reliability,* Networks 23, 1- 17, 1993.

[Cayley, 1889] A. Cayley, *A theorem on trees,* Quart. J. Math 23,376-378: 69, 1889.

[Cheng, 1981] C. Cheng, *Maximizing the total number of spanning trees in a graph: two related problems in graph theory and optimum design theory,* Journal of combinatorial Theory B 31,1981.

[Chartrand, 2011] G. Chartrand, L. Lesniak, and P. Zhang, *Graphs and Digraphs (5th ed),* CRC Press, Boca Raton, FL, 2011.

[Erdos, 1960] P. Erdos and T. Gallai, *Graphs with prescribed degrees of vertices* (Hungarian), Mat. Lapok 11, 264-274, 1960.

[Fiedler, 1973] M. Fieldler, *Algebraic Connectivity of Graphs,* Czech. Math Journal 23, 298-305, 1973.

[Fuller, 2010] J. Fuller. *Spanning trees of extended ideal proper split graphs.* Honors Thesis, Seton Hall University, Department of Mathematics and Computer Science, 2010.

[Hakimi, 1962] Hakimi, S. L. On realizability of a set of integers as degrees of the vertices of a linear graph. I. *J. Soc. Indust. Appl. Math.* 10, 496–506, 1962.

[Hammer, 1978] P.L. Hammer, T. Ibaraki, and B. Simeone. *Degree sequences of threshold graphs.* Congressus Numeratium, Vol. 21, 329-355, 1978.

[Hammer, 1981] P.L. Hammer and B. Simeone. *The splittance of a graph.* Combinatoria Vol. 1, 275-284, 1981.

[Heinig, 2012] M. Heinig and J.T. Saccoman, *Laplacian integral multigraphs,* Congressus Numerantium 212, 131-143, 2012.

[Kelmans, 1967] A.K. Kelmans, *Properties of characteristic polynomial of a graph,* Kibernetiku-na sluzbu kommunizmu 4, Energija, Moskva-Leningrad, 27-41, 1962.

[Kelmans, 1974] A.K. Kelmans and V. M. Chelnokov, *A certain polynomial of graph and graphs with an extremal number of trees,* Journal of Combinatorial Theory B 16, 197-214, 1974.

[Kelmans 1981]. *On graphs with randomly deleted edges,* Acta Math Acad. Sci Hungary 37, 77-88, 1981.

[Kirchhoff, 1847] *Über die ausflösung der gleichungen, auf welche man bei der untersuchung der linearen vertheilung galvanischer ströme geführt wird. ("On the solution of the equations obtained from the investigation of the linear distribution of galvanic currents").* Annalen der Physik und Chemie 72,497-508, 1847.

[Lancaster, 1985] P. Lancaster and M. Tismenetsky, *Theory of Matrices,* Academic Press, Orlando, FL, 1985.

[Marshall, 1979]. A.W. Marshall and L. Olkin, *Inequalities: The Theory of Majorization and its Applications,* Academic Press, New York, 1979.

[Merris, 1994] R. Merris, *Degree maximal graphs are Laplacian integral,* Linear Algebra Appl. 199, 381–389, 1994.

[Michewicz, 2007] J. Michewicz and J.T. Saccoman. *A formula for the number of spanning trees of certain non-threshold split graphs.* Congressus Numerantium, Vol. 184, 85-96, 2007.

[Moore, 2013] K. Moore and J.T. Saccoman. *Spanning trees of almost ideal proper split graphs.* Congressus Numerantium, Vol. 216, 129-141, 2013.

[Moskowitz, 1958] F. Moskowitz, *The analysis of redundancy networks,* AIEE Transactions on Commumication Electronics 39, 627-632, 1958.

[Petingi, 1996] L. Petingi, J.T. Saccoman. and L. Schoppmann, *Uniformly least reliable graphs,* Networks 27(2), 125-131, 1996.

[Seshu, 1961] S. Seshu and M. Reed, *Linear Graphs and Electrical Networks,* Addison-Wesley, Reading, MA, 1961.

[Shier, 1974] D. Shier, *Maximizing the spanning trees in a graph with n nodes with n nodes and m edges,* Journal of Research of the National Bureau of Standards-B 78B, 193-196, 1974.

[Schoppmann, 1990] L. Schoppmann, *Network bounds for edge reliability,* Doctoral dissertation, Stevens Tech, Mathematics Department, 1990.

[Satyanarayana, 1992] A. Satyanarayana, L. Schoppmann, and C.L. Suffel, *A reliabilty-improving graph transformation with applications to network reliability,* Networks 22, 209-216, 1992.

[Temperley, 1964] H.N.V. Temperley, *On the mutual cancellation of cluster integrals in Mayer's fugacity series, Proc. Phys. Soc.* **83**, 3–16, 1964.

Index

Printed in the United States
By Bookmasters